SPACE SCIENCE, EXPLORATION AND POLICIES



# HUMAN SPACE EXPLORATION

## EARLY ASSESSMENTS OF NASA'S NEXT STEPS

# SPACE SCIENCE, EXPLORATION AND POLICIES

Additional books in this series can be found on Nova's website under the Series tab.

Additional e-books in this series can be found on Nova's website under the e-book tab.

SPACE SCIENCE, EXPLORATION AND POLICIES

# HUMAN SPACE EXPLORATION

## EARLY ASSESSMENTS OF NASA'S NEXT STEPS

**DARREL GIBBS**
**EDITOR**

NOVINKA

*New York*

**Library of Congress Cataloging-in-Publication Data**

ISBN: 978-1-63482-172-8

*Published by Nova Science Publishers, Inc. † New York*

SPACE SCIENCE, EXPLORATION AND POLICIES

# HUMAN SPACE EXPLORATION

## EARLY ASSESSMENTS OF NASA'S NEXT STEPS

DARREL GIBBS
EDITOR

NOVINKA
*New York*

## NOTICE TO THE READER



Additional color graphics may be available in the e-book version of this book.

**Library of Congress Cataloging-in-Publication Data**

ISBN: 978-1-63482-172-8

*Published by Nova Science Publishers, Inc. † New York*

# CONTENTS

# PREFACE

This book examines the scope of NASA's preliminary cost estimates for the three programs. It examines the SLS program's progress toward and risks for its first test flight in 2017; and the extent to which the SLS program has plans in place to achieve its long-term goals and promote affordability.

Chapter 1 - NASA is undertaking a trio of closely-related programs to continue human space exploration beyond low-Earth orbit: the SLS vehicle; the Orion capsule, which will launch atop the SLS and carry astronauts; and the supporting ground systems.

As a whole, the efforts represent NASA's largest exploration investment over the next decade, potentially as much as $22 billion, to demonstrate initial capabilities. Beyond 2021, NASA plans to incrementally develop progressively more-capable SLS launch vehicles complemented by Orion capsules and ground systems.

GAO was asked to assess the costs of NASA's human exploration program. This report examines the scope of NASA's preliminary cost estimates for the three programs.

To conduct this work, GAO reviewed NASA information on cost estimates for the three programs, discussed the estimates with NASA officials, and assessed the estimates against best practices criteria in GAO's cost estimating guidebook as well as NASA's own requirements and guidance.

Chapter 2 - SLS is NASA's first exploration-class heavy lift launch vehicle in over 40 years. Predecessor programs, such as Constellation, were canceled in the face of acquisition problems and funding shortfalls. NASA estimates it could spend almost $12 billion developing the first of three SLS vehicle variants and associated ground systems through initial launch in late 2017 and potentially billions more to develop increasingly capable vehicles.

Ensuring that this program is affordable and sustainable for the long term is a key goal of the 2013 National Space Transportation Policy.

GAO was asked to evaluate SLS program challenges. This report examines (1) the SLS program's progress toward and risks for its first test flight in 2017 and (2) the extent to which the SLS program has plans in place to achieve its long-term goals and promote affordability. To do this, GAO reviewed relevant design, development, cost, and schedule documents; interviewed program officials; and evaluated SLS program actions using acquisition and cost estimating best practices.

Chapter 3 - Statement of Cristina T. Chaplain, Director, Acquisition and Sourcing Management, Government Accountability Office.

Chapter 4 - Statement of William H. Gerstenmaier, Associate Administrator for Human Exploration and Operations, NASA.

In: Human Space Exploration
Editor: Darrel Gibbs

ISBN: 978-1-63482-172-8


*Chapter 1*

# NASA: Actions Needed to Improve Transparency and Assess Long-Term Affordability of Human Exploration Programs*

*United States Government Accountability Office*

## Why GAO Did This Study

NASA is undertaking a trio of closely-related programs to continue human space exploration beyond low-Earth orbit: the SLS vehicle; the Orion capsule, which will launch atop the SLS and carry astronauts; and the supporting ground systems.

As a whole, the efforts represent NASA's largest exploration investment over the next decade, potentially as much as $22 billion, to demonstrate initial capabilities. Beyond 2021, NASA plans to incrementally develop progressively more-capable SLS launch vehicles complemented by Orion capsules and ground systems.

GAO was asked to assess the costs of NASA's human exploration program. This report examines the scope of NASA's preliminary cost estimates for the three programs.

* This is an edited, reformatted and augmented version of a United States Government Accountability Office publication, No. GAO-14-385, dated May 2014.

To conduct this work, GAO reviewed NASA information on cost estimates for the three programs, discussed the estimates with NASA officials, and assessed the estimates against best practices criteria in GAO's cost estimating guidebook as well as NASA's own requirements and guidance.

## WHAT GAO RECOMMENDS

NASA should establish separate cost baselines that address the life cycle of each SLS increment, as well as for any evolved Orion or ground systems capability, among other actions to enable assessment of affordability and enhance oversight.

In commenting on a draft of this report, NASA partially concurred with GAO's recommendations, citing that actions taken to structure the programs and track costs met their intent. However, GAO believes NASA's responses do not fully address the issues raised in this report.

## WHAT GAO FOUND

The scope of the National Aeronautics and Space Administration's (NASA) preliminary cost estimates for the Space Launch System (SLS), Orion Multi-Purpose Crew Vehicle (Orion), and associated ground systems encompasses only the programs' initial capabilities and does not include the long-term, life cycle costs associated with the programs or significant prior costs:

- The SLS estimate is based on the funding required to develop and operate the initial 70-metric ton variant through first flight in 2017 but not the costs for its second flight in 2021. NASA is now incurring some costs related to the second flight, but it is not currently tracking those costs for life cycle cost estimating purposes. Furthermore, the estimate does not include costs to incrementally design, develop, and produce future 105- and 130-metric ton SLS variants which NASA expects to use for decades. NASA is now funding concept development and analysis related to these capabilities.
- The Orion estimate does not include costs for production, operations, or sustainment of additional crew capsules, despite plans to use and

possibly enhance this capsule after 2021. It also does not include $4.7 billion in prior costs incurred during the approximately 4 years when Orion was being developed as part of NASA's now-defunct Constellation program.

- The ground systems estimate excludes costs to develop or operate the ground systems infrastructure beyond 2017, although NASA intends to modify ground architecture to accommodate all SLS variants.

NASA expects to use this same limited scope of work to establish the programs' baseline cost estimates in 2014. According to NASA, the agency is developing a tailored definition for the programs' life cycle cost estimates as allowed by NASA requirements. Agency officials stated that NASA chose its approach in part due to uncertainties about the programs' end dates and missions beyond 2021.

GAO recognizes that defining life cycle costs can be difficult when uncertainties exist, and that best practices for cost estimating look favorably on evolutionary development. Even so, best practices expect that a high-quality cost estimate will account for program uncertainties, forecast a minimum and maximum range for all life cycle costs, and clearly define the characteristics of each increment of capability so that a rigorous life cycle cost estimate can be developed.

According to these practices as well as NASA's requirements and guidance, life cycle cost estimates should encompass all past, present, and future costs for a program, including costs for operations, support, and disposal. The limited scope that the agency has chosen for constructing preliminary and baseline cost estimates, however, means that the estimates are unlikely to serve as a way to measure progress and track cost growth over the life of the programs. For example, cost growth on the current SLS variant could be masked as the addition of scope associated with work for future variants, and the baseline estimate would no longer be applicable. Insight into program costs helps decision makers understand the long-term affordability of programs—a key goal of the National Space Transportation Policy—and helps NASA assess management of its portfolio to achieve increasing capabilities as directed in the NASA Authorization Act of 2010.

May 8, 2014

The Honorable Tom Coburn, M.D.
Ranking Member
Committee on Homeland Security and Governmental Affairs
United States Senate

Dear Senator Coburn:

The National Aeronautics and Space Administration's (NASA) Human Exploration and Operations Mission Directorate (HEO) is undertaking a trio of closely-related programs to continue human space exploration beyond low-Earth orbit. These programs include the Space Launch System (SLS) vehicle to launch astronauts and carry cargo into space, the Orion Multi-Purpose Crew Vehicle (Orion) capsule that will launch atop the SLS and carry the astronauts, and the Ground Systems Development and Operations (GSDO) to support such activities as assembly, test, and launch of the SLS and Orion. As a whole, the efforts represent NASA's largest exploration investment over the next decade and may cost as much as $22 billion to demonstrate initial capabilities that encompass the first SLS flight in 2017, the ground systems for that effort, and the first two Orion flights in 2017 and 2021.

GAO has designated NASA's management of acquisitions as a high-risk area for more than two decades in view of persistent cost growth and schedule slippage in the majority of its major projects. Last year we reported that the agency had made progress in reducing risk on smaller-scale, less complex projects after years of struggling with poor cost estimation, weak oversight, and risk underestimation. We also reported, however, that demonstrating that this progress can be translated to larger, more complex projects, such as SLS and Orion, will be especially important in an era of constrained budgets and competing priorities.[1] Establishing an exploration program that will be affordable and sustainable for the long-term is also a key goal of the 2013 National Space Transportation Policy.[2]

In light of these issues, you requested that we review the costs of NASA's exploration programs. Thus, we examined the agency's preliminary cost estimates for the SLS, Orion, and GSDO programs. Specifically, we identified the scope of the cost estimates and assessed whether that scope provided transparency into costs and enabled assessment of longterm affordability. This report provides the results of our review. In other on-going work, we are performing a more in-depth assessment addressing the credibility and reliability of NASA's cost estimate for SLS.

To conduct this work, we reviewed NASA's preliminary cost estimates for the SLS, Orion, and associated ground systems programs and information related to the baseline cost estimates for the programs in order to determine the scope of the estimates. We assessed the estimates' scope against best practices criteria outlined in GAO's cost estimating guidebook.[3] We also discussed the estimates with NASA officials, including the rationale for the estimates' scope and exclusions to that scope. We did not asses the reliability of the SLS, Orion, and GSDO cost estimates.

We conducted this performance audit from March 2014 through May 2014 in accordance with generally accepted government auditing standards. Those standards require that we plan and perform the audit to obtain sufficient, appropriate evidence to provide a reasonable basis for our findings and conclusions based on our audit objectives. We believe that the evidence obtained provides a reasonable basis for our findings and conclusions based on our objectives.

## BACKGROUND

The National Aeronautics and Space Administration Authorization Act of 2010 directed NASA to develop a Space Launch System as a follow-on to the Space Shuttle and as a key component in expanding human presence beyond low-earth orbit. The Act also directed NASA to continue development of a multi-purpose crew vehicle for use with that system.[4] To that end, NASA plans to incrementally develop three progressively-larger SLS launch vehicle capabilities—70-, 105- and 130-metric ton (MT) variants—complemented by the Orion and supporting ground systems.

Figure 1 below illustrates NASA's planned capabilities for the SLS, Orion, and some of the related GSDO efforts.

These capabilities follow the agency's previous attempt to develop a next-generation human spaceflight system, the Constellation program, which was cancelled in 2010 when the program's budget proved inadequate to resolve technical challenges. The first version of the SLS being developed is a 70-metric ton launch vehicle known as Block I. NASA expects to conduct two test flights of the Block I vehicle—the first in 2017 and the second in 2021. The vehicle is scheduled to fly some 700,000 kilometers beyond the moon during the first test flight, known as Exploration Mission-1 (EM-1), and to fly a second mission, known as Exploration Mission-2 (EM-2), to test additional aspects of its performance. After 2021, NASA intends to build 105- and 130-metric ton launch vehicles, known respectively as Block IA/B and Block II, which it expects to use as the backbone of manned spaceflight for decades. NASA anticipates that these launch vehicles will require the development of new systems to achieve the agency's goals for carrying greater amounts of cargo and traveling farther into space. The agency has not yet selected specific missions for the increased capabilities to be provided by Block IA/B and Block II but, in keeping with the language contained in the 2010 Authorization Act, anticipates using the vehicles for such deep-space destinations as near-Earth asteroids and Mars.

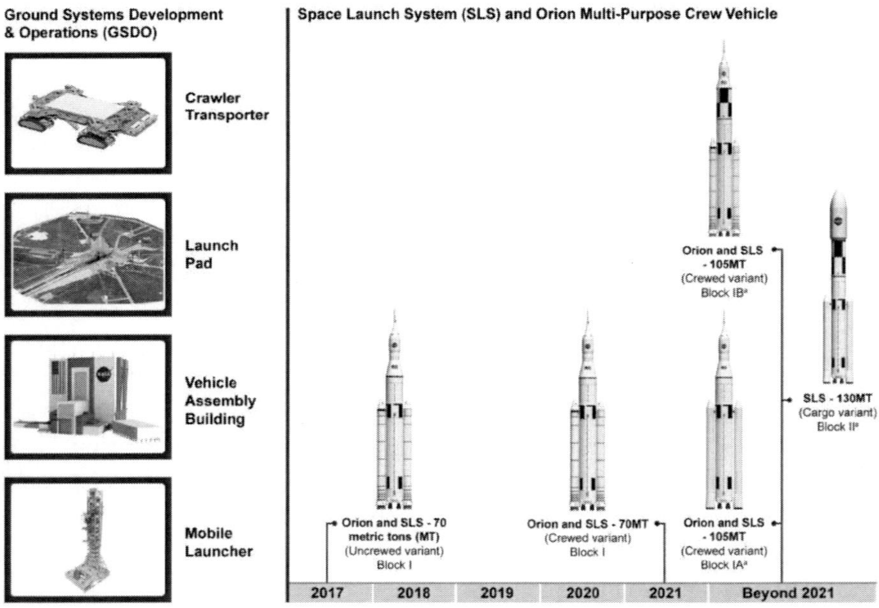


Source: NASA (data and images).

[a] NASA plans for SLS Block IA to utilize advanced boosters, Block IB an exploration upper stage, and Block II the advanced boosters and exploration upper stage. The agency has not yet determined whether it will first develop the Block IA or Block IB variant.

Figure 1. SLS, Orion, and GSDO Capabilities.

In concert with SLS, NASA expects to evolve the Orion and ground systems. The agency plans an un-crewed Orion capsule to fly atop the SLS during EM-1 in 2017, a crewed capsule during EM-2 in 2021, and ultimately, at a date to be determined, a crewed capsule with capability for such missions as a Mars landing. NASA is also modifying the existing ground systems so that they can support the SLS Block I variant and eventually accommodate the Block IA/B and Block II launch vehicles as well as enhanced versions of the Orion crew capsule. For example, NASA plans to add moveable floors to the vehicle assembly building at Kennedy Space Center so that the three launch vehicle variants can be more easily prepared for flight as the SLS capability evolves.

NASA established the preliminary cost estimates for the initial capabilities of the SLS, Orion, and associated GSDO as each of these programs entered the preliminary design and technology completion phase of development, known as key decision point B (KDP-B).[5] At KDP-B, programs use a probability-based analysis to develop a range of preliminary cost and schedule estimates which are used to inform the budget planning for the programs. This phase culminates in a review at key decision point C (KDP-C), known as program confirmation, where cost and schedule baselines with point estimates are established and documented in the agency baseline commitment.[6] After this review, programs are considered to be in the implementation phase of development, and program progress is subsequently measured against these baselines.[7] NASA plans to hold the program confirmation review for SLS in spring 2014 and expects to conduct the KDP-C review for GSDO in May 2014 and Orion in December 2014. Because the life cycle costs of these programs are expected to exceed $250 million, NASA is required to report the programs' baseline estimates to Congress once the programs are approved to move into implementation.[8] The agency provides this information through its annual budget submission. NASA also uses the annual budget submission to inform Congress about the preliminary cost ranges for projects proceeding into formulation.

## LIMITED SCOPE OF EXPLORATION PROGRAM ESTIMATES DOES NOT CAPTURE LIFE CYCLE COSTS

NASA's preliminary cost estimates for the SLS, Orion, and associated GSDO programs do not provide a complete picture of the costs required to develop and operate the programs through the entire course of their respective life cycles.

These preliminary estimates include the funding required for the scope of work related to initial capabilities—that is, development and operations through 2017 for the SLS launch vehicle and ground systems and through 2021 for the Orion.

NASA also expects to use this same limited scope of work to develop the SLS, Orion, and GSDO baseline cost estimates. Moreover, NASA's estimates do not capture the cost of the second flight of the 70-metric ton vehicle during EM-2, the costs of development work that will be necessary to fly the increased 105- and 130-metric ton SLS capabilities, and the costs associated with legacy hardware that will be used for the Orion program. In contrast, best practices for cost estimation call for "cradle to grave" life cycle cost estimates in order to help assess a program's long-term affordability.

## Cost for Initial SLS, Orion, and GSDO Capabilities Totals up to $22 Billion

NASA's preliminary cost estimates for the three programs' initial capabilities total a low-to-high cost range of approximately $19 to $22 billion. Table 1 below depicts the scope, including content and schedule, of the SLS, Orion, and GSDO initial capabilities' preliminary cost estimates. As the SLS, Orion, and GSDO programs move from formulation into implementation phases, NASA plans to use the same content and scope for calculating the programs' respective baseline cost estimates.

## NASA's Estimates Do Not Include Life Cycle Costs

NASA's preliminary cost estimates for SLS, Orion, and GSDO provide no information about the longer-term, life cycle costs of developing, manufacturing, and operating the launch vehicle, crew capsule, and ground systems:

**Table 1. Scope of the SLS, Orion, and GSDO Preliminary Cost Estimates**

| | | | | Preliminary Cost Estimate (Fiscal Years 2012-2018) (Dollars in Billions) | | |
|---|---|---|---|---|---|---|
| **Capability** | **Time Frame** | **Associated Flight** | **Content** | **Minimum** | **Maximum** | **AnticipatedDate forBaseline CostEstimate** |
| Space Launch System (SLS) | Flight in December 2017 | Exploration Mission-1 (EM-1) | • Design, develop,and build 70-metric ton launch vehicle<br>• Establish manufacturing facilities for that vehicle<br>• Conduct un-crewed first exploration flight in 2017<br>• Perform 3 months of post-flight data analysis | $7.7 | $8.6 | May 2014 |
| Ground Systems Development and Operations (GSDO) | Flight in December 2017 | EM-1 | • Develop the ground systems infrastructure for assembly, test, launch, and recovery associated with EM-1<br>• Perform 3 months of post-flight Orion recovery and operations | $2.8 | $3.1 | May 2014 |
| Orion Multi-Purpose Crew Vehicle (Orion) | Flights in December 2017 and August 2021 | EM-1 and EM-2 | • Development, manufacturing, and operations for the un-crewed Orion flight on SLS in 2017 and crewed Orion flight on SLS in 2021<br>• Perform 3 months of post-flight data analysis | $8.5 | $10.3 | December 2014 |
| **Total Estimated Costs** | | | | **$19.0** | **$22.0** | |

Source: GAO analysis of NASA documentation.

- The SLS estimate does not cover the cost to build the second 70-metric ton vehicle and conduct EM-2 in 2021 with that vehicle. NASA is already incurring costs for EM-2 because it is funding some EM-2 development in concert with EM-1 efforts, such as work on the solid rocket boosters and core stage that are expected to help power the 70-metric ton SLS. NASA officials indicated at one point in our review that they did not expect to begin formally tracking EM-2 costs until after the SLS design's maturity was assessed at a critical design review scheduled for 2015; however, the agency stated in technical comments to this report that it is tracking those costs for budget purposes and plans to begin formally reporting them once SLS reaches the project confirmation phase. Additionally, the SLS estimate does not address the potential for costs NASA would incur to produce, operate, and sustain flights of the 70-MT Block I capability beyond 2021. NASA officials stated that there are currently no plans to fly that vehicle beyond 2021, but that the agency could reassess its decision if a specific mission arises for the vehicle.
- The SLS estimate also does not include costs to design, develop, build, and produce the 105- or 130-metric ton Block IA/B and Block II SLS variants that NASA intends to use well into the future. NASA indicated that these variants will require new systems development efforts—including advanced boosters and a new upper stage to meet the greater performance requirements associated with larger payloads as well as travel to Mars or other deep-space locations. NASA has started funding concept development, trades, and analyses related to these new designs, such as assessing the use of lightweight materials to construct the upper stage and selective laser melting to produce system components. In addition, NASA anticipates a re-start of the production line for the RS-25 engine that it plans to use to power the Block IA/B and Block II vehicles. Currently, the agency has enough residual RS-25 liquid-fuel engines from the Space Shuttle program to launch the SLS for up to 4 flights. NASA expects to need more of the engines beyond that, but it has not yet finalized acquisition plans to manufacture them. According to agency officials, re-starting the production line would entail at least 3 years, whereas development of a new engine would require a minimum of 8 years.
- The Orion estimate does not address costs for production, operations, or sustainment of additional crew capsules after 2021 nor does it address prior costs incurred when Orion was being developed as part

of the now-defunct Constellation program. NASA initiated the crew capsule's development in 2006 as part of the Constellation program. During approximately 4 years that the capsule's development occurred under Constellation, the agency spent about $4.7 billion for the capsule's design and development. When Constellation was cancelled in 2010 and the work transitioned to the current Orion program, however, NASA excluded the Constellation-related costs from Orion's current preliminary cost estimate of $8.5 to $10.3 billion through 2021.

- The GSDO estimate does not address the costs to develop or operate SLS ground systems infrastructure beyond EM-1 in 2017, although NASA intends to modify ground architecture to accommodate all SLS variants.

NASA officials have indicated that the road ahead involves many decisions about the programs beyond 2021, including how development will proceed, what missions will be performed, when the programs will end, and how each effort will be managed. They noted that the agency is using a capability-based approach to SLS, Orion, and the associated GSDO development, in which system capability grows over time. They indicated that the programs' preliminary cost estimates are for attainment of capabilities rather than the full cost of the programs, and that it is difficult to define life cycle costs because the programs' intended longterm uses and life spans have not been fully determined. According to NASA, the agency is developing a tailored definition for life cycle cost estimating that is allowed by NASA requirements. Because the missions drive the number and types of vehicles, crew capsules, and ground systems that would be required, as missions are defined, NASA officials said they would be in a better position to estimate the programs' life cycle costs. The officials stated that NASA is looking ahead to future costs as much as possible, and NASA indicated in technical comments to this report that the SLS program plans to begin formally reporting costs for the launch vehicle's EM-2 after the program's anticipated confirmation in spring 2014.

We recognize that defining life cycle costs can be difficult when uncertainties exist. However, in contrast to NASA's tailored approach, both widely-accepted best practices for cost estimation and the agency's own requirements support the need for full life cycle cost estimates. Even when uncertainties exist, best practices maintain that a high-quality cost estimate takes into account those uncertainties while forecasting the minimum and maximum range of all life cycle costs. The best practices, developed by the GAO in concert with the public and private sector cost estimating communities, call for "cradle to grave" life cycle cost estimates and maintain that life cycle cost estimates should provide an exhaustive, structured accounting of all resources and associated cost elements required to develop, produce, deploy, and sustain a particular program. This entails identification of all pertinent cost elements, from initial concept through operations, support, and disposal. Likewise, NASA's program management requirements direct that programs develop a preliminary full life cycle cost estimate. In accordance with the agency's guidance regarding life cycle costs, such an estimate would encompass total costs from the formulation through the implementation phase, including design, development, mission operations, support, and disposal activities.

According to best practices, because life cycle estimates encompass all possible costs, they provide a wealth of information about how much programs are expected to cost over time. Life cycle cost estimates, including a range for preliminary costs as directed by NASA requirements for programs in the formulation phase, enhance decision making, especially in early planning and concept formulation of acquisition. High-quality cost estimates, as noted by best practices, can support budgetary decisions, key decision points, milestone reviews, and investment decisions. For example, a preliminary life cycle cost estimate provides the basis of the financial investment that the agency is committing the government to, while a baseline life cycle cost estimate forms the basis for measuring cost growth over time. Because NASA expects to continue with a limited scope for the SLS, Orion, and baseline estimates, however, cost growth over time within the programs will be difficult to identify and could be masked as growth in the SLS capability if the most current cost estimate did not contain the same content as the baseline estimate.

As noted in best practices for cost estimating, the quality of a program's cost estimate is also key to determining its affordability, that is, the degree to which a program's funding requirements fit within an agency's overall portfolio plan. However, NASA's preliminary cost estimates do not address the affordability of increased capabilities because they exclude the life cycle costs associated with the SLS Block IA/B and Block II launch vehicles that the agency intends to use well into the future. According to agency officials at the time of our review, NASA has not yet decided whether it will manage the Block IA/B and Block II development efforts as individual programs and, if so, what the programs' scope would be. Best practices for cost estimating look favorably on the incremental development approach NASA has chosen for

SLS, and they also state that programs following such an approach should clearly define the characteristics of each increment of capability so that a rigorous life cycle cost estimate can be developed. In addition, we have previously concluded that it is prudent for an agency to manage increasing capabilities of an existing program on par with the investments yet to come and in a way that is beneficial for oversight. For example, we have recommended that agencies developing weapon systems in increments consider establishing each increment of increased capability with its own cost and schedule baseline.[9] According to cost estimating best practices, dividing programs into smaller pieces makes management and testing easier and helps avoid unrealistic cost estimates, resulting in more realistic long-range investment funding and more effective resource allocation. These are important considerations given that NASA is likely to spend billions of dollars beyond its initial investment of up to $22 billion to develop the increased capabilities. Development of human-rated liquid-fueled engines, for example, has been among the most difficult, time-intensive, and costly parts of launch vehicle development. As a case in point, NASA spent about 8 years and $1.5 billion to develop a human-rated engine known as J-2X for use on Ares launch vehicles within the agency's now-defunct Constellation program.

NASA has faced issues with affordability of its manned space flight investments and other major projects in the past, and those affordability issues have sometimes contributed to a program's cancellation. For example, NASA originally envisioned that the Space Shuttle would fly up to 100 times per vehicle at a cost of $7.7 million per launch.[10] In reality, the Shuttle flew 135 times in total over a period of 30 years at a cost that was about $3.5 billion per year around the 2008 timeframe. Amid concerns that included the Shuttle's costs and safety, the program ended. NASA then focused on building human spaceflight alternatives that included Constellation. In 2010, Constellation was canceled because, as noted by NASA's Administrator, the program could not return astronauts to the moon at an affordable cost and would require far more funding to make the agency's approach viable.[11] In a recent example noted in the agency's 2015 presidential budget request, NASA may place in storage the Stratospheric Observatory for Infrared Astronomy, an airborne observatory for studying astronomical objects and phenomena, after spending some 23 years and more than $1 billion to develop the project. The agency cited high operating costs, estimated at some $1.8 billion over the project's planned life, as a factor in its considerations.

## CONCLUSION

The SLS, Orion, and GSDO programs NASA has established to fulfill its mandate of providing the capability for transporting humans to space are well underway. These programs represent a significant investment for the country—as much as $22 billion for initial capabilities and potentially billions more to field increased capabilities over time as envisioned in the 2010 NASA Authorization Act. Given the goals that have been outlined for NASA as part of the National Space Transportation Policy, the success of these programs is to be measured not only by the capability that is achieved but also by NASA's ability to achieve them within a reasonable timeframe and cost to the U.S. taxpayer. As such, establishing these programs with both near-term and long-term affordability in mind is key. The limited scope that NASA has chosen to use as the basis for formulating the programs' cost baselines, however, does not provide the transparency necessary to assess long-term affordability and will hamper oversight by those tasked with assessing whether the agency is progressing in a cost-effective and affordable manner. If the SLS, Orion, and GSDO baseline cost estimates cannot be compared to current costs, the baseline estimates lose their usefulness because they no longer serve as a means to hold NASA accountable for cost growth and program progress. Furthermore, if NASA does not clearly delineate costs for operations and sustainment of the initial capabilities or separate cost and schedule baselines for upcoming capabilities, then it will be difficult to assess program affordability and for the Congress to make informed, long-term budgetary decisions. Estimates that use all available information to establish a potential range of costs for the full scope of these upcoming capabilities can help inform such decisions.

## RECOMMENDATIONS FOR EXECUTIVE ACTION

To provide the Congress with the necessary insight into program affordability, ensure its ability to effectively monitor total program costs and execution, and to facilitate investment decisions, we recommend that NASA's Administrator direct the Human Exploration and Operations Mission Directorate take the following 3 actions:

- Establish a separate cost and schedule baseline for work required to support the SLS Block I EM-2 and report this information to the Congress through NASA's annual budget submission. If NASA decides to fly the SLS Block I beyond EM-2, establish separate life cycle cost and schedule baseline estimates for those efforts, to include funding for operations and sustainment, and report this information annually to Congress via the agency's budget submission.
- Because NASA intends to use the increased capabilities of the SLS, Orion, and GSDO efforts well into the future and has chosen to estimate costs associated with achieving the capabilities, establish separate cost and schedule baselines for each additional capability that encompass all life cycle costs, to include operations and sustainment. When NASA cannot fully specify costs due to lack of well-defined missions or flight manifests, forecast a cost estimate range — including life cycle costs — having minimum and maximum boundaries. These baselines or ranges should be reported to Congress annually via the agency's budget submission.
- Because a significant amount of the original Orion development work occurred under the Constellation program, include those costs in the baseline cost estimate for the Orion program.

## AGENCY COMMENTS AND OUR EVALUATION

In responding to a draft of our report, NASA partially concurred with our three recommendations, citing among other reasons that actions already in place at the time of our review such as establishing SLS, Orion, GSDO as separate programs and a block upgrade approach for SLS—and actions it plans to take to track costs—met the intent of our recommendations. In most cases, the actions that NASA plans to take do not fully address the issues we raised in this report. We continue to believe that our recommendations are valid and should be fully addressed as discussed below. NASA also provided technical comments which we incorporated as appropriate.

NASA partially concurred with our first recommendation to establish a separate cost and schedule baseline for work required to support the SLS Block I EM-2, report this information to the Congress through NASA's annual budget submission, and establish separate life cycle cost and schedule baseline estimates for EM-2 if NASA decides to fly Block I beyond EM-2. NASA also partially concurred with our second recommendation to establish separate cost and schedule baselines that encompass life cycle costs, including operations and sustainment, for each additional SLS, Orion, and GSDO capability and to report cost estimates for the capabilities annually via the agency budget submission until key requirements are defined and baselines can be established. In its response, NASA stated that it had established separate programs for SLS, Orion, and GSDO and adopted a block upgrade approach for SLS. This approach, NASA stated, is in concert with best practices and NASA policy. In addition, NASA indicated that it will establish cost and schedule estimates for initial demonstration of the three programs as they enter respective implementation phases and will begin reporting development, operations, and sustainment costs for SLS Block I and subsequent variants starting in fiscal year 2016 via its annual budget submission to Congress. Finally, the agency stated that it intends to conduct design reviews for upgraded SLS elements, including the upper stage and booster, and set up cost commitments similar to what it has done for Block I capability as part of that design review process, but that it does not intend to establish life cycle estimates for SLS through the end of the program because flight rates, mission destinations and other strategic parameters are yet unknown.

As discussed in the report, best practices for cost estimating recognize that NASA's evolutionary development approach for SLS, Orion, and GSDO helps reduce risk and provide capabilities more quickly. Given NASA's planned long-term use of the SLS, Orion, and GSDO, its block upgrade approach and intention to conduct design reviews for each of the planned upgrades will provide some understanding of the development work and resources required. For example, such reviews are typically expected to yield information about technical progress against requirements. While NASA's prior establishment of SLS, Orion, and GSDO as separate programs lends some insight into expected costs and schedule at the broader program level, it does not meet the intent of our first two recommendations because cost and schedule identified at that level is unlikely to provide the detail necessary to monitor the progress of each block against a baseline. Furthermore, it is unclear from NASA's response whether the cost commitments the agency plans within the design review process will serve the same purpose as establishing a cost baseline for each respective upgrade.

Additionally, NASA's planned approach for reporting costs associated with EM-2 and subsequent variants of SLS via its annual budget submission only partially meets the intent of our first two recommendations. Providing cost information at an early phase when baseline estimates have yet to be established is helpful to ensure costs associated with EM-1 and EM-2 are not conflated and funding requirements for future flights of the Block I SLS and future variants are somewhat understood. Reporting the costs via the budget process alone, however, will not provide information about potential costs over the longterm because budget requests neither offer all the same information as life cycle cost estimates nor serve the same purpose. Plainly, progress cannot be assessed without a baseline that serves as a means to compare current costs against expected costs. An agency's budget submission reflects its current annual fiscal needs and anticipated short-term needs up through an additional 4-year period for a particular program, is subject to change based on fiscal negotiation, and is not necessarily linked to an established baseline that indicates how much the agency expects to invest to develop, operate, and sustain a capability over the long-term. Conversely, life cycle cost estimates establish a full accounting of all program costs for planning, procurement, operations and maintenance, and disposal and provide a long-term means to measure progress over a program's life span. As NASA establishes parameters for the additional flights of the first SLS capability and upgraded capabilities, including flight rates, mission destinations, and other requirements, it will be well-poised to move from reporting costs in budget submissions to establishing baseline cost and schedule estimates for each capability and reporting progress against these respective baselines. Therefore, we continue to believe that NASA should baseline costs for EM-2 and each future variant of SLS and report progress against those established baselines.

NASA makes no specific mention of how it plans to account for future work associated with Orion and GSDO. We believe it is important to treat Orion and GSDO with the same significance as SLS because this trio of programs is expected to work in concert now and in the future to achieve NASA's goals for human space exploration. Reporting Orion and GSDO development, operations, and sustainment costs in the annual budget request, as NASA plans for SLS, would be a logical first step. Just as with SLS, however, it will be important for NASA to establish and report progress against baseline costs and schedules for each block of Orion and GSDO efforts as flight rates, missions, and other strategic parameters are defined because doing so will help the agency more effectively manage not only each program but its human exploration portfolio as a whole.

NASA partially concurred with our third recommendation to include the costs of Orion development work under the Constellation program as part of the baseline cost estimate for the Orion program. Agency officials stated that they agree those costs should be tracked and disclosed, but that the current

Orion program has a new concept of operations, requirements, and budget plan than that under the Constellation effort.

The past costs incurred for Orion's development are important because they provide visibility into the total cost of developing a crew capsule for human space exploration. Exclusion of these costs from Orion's current estimate understates how much NASA will invest to put humans into space. Although NASA notes that it has changed Orion's concept of operations and requirements, the agency nonetheless migrated Orion critical technology development efforts from Constellation to the SLS program. For example, NASA began efforts to develop the coating for Orion's heat shield as part of Constellation, and the agency continues that development today in preparation for the capsule's launch atop SLS. Therefore, we continue to believe our recommendation to include Orion development costs under Constellation in the baseline cost estimate for the current Orion program is valid and should be fully implemented.

Sincerely yours,
Cristina T. Chaplain Director
Acquisition and Sourcing Management

## End Notes

[1] GAO, *NASA: Assessments of Large-Scale Selected Projects*, GAO-13-276SP, (Washington, D.C.: May 17, 2013).

[2] National Space Transportation Policy, November 21, 2013.

[3] GAO, *GAO Cost Estimating and Assessment Guide: Best Practices for Developing and Managing Capital Program Costs,* GAO-09-3SP, (Washington, D.C.: March 2009).

[4] Pub. L. No. 111-267, §§ 302(a), 303(a) (codified at 42 U.S.C. §§ 18322, 18323).

[5] KDP-B occurs during a program's formulation phase. The formulation phase is the first phase of a NASA program, wherein requirements are defined (i.e., what the program is designed to do), schedules and cost estimates established, and implementation plans produced.

[6] NASA defines baseline as the technical performance and content, technology application, schedule milestones, and budget (including contingency and allowance for program adjustment) that are documented in the approved program plans.

[7] In the implementation phase, a program typically carries out plans made in formulation, performs final design and fabrication, tests components and system assembly, integrates these components and tests how the components work together, and launches on its intended mission(s).

[8] Under the National Aeronautics and Space Administration Authorization Act of 2005, Pub. L. No. 109-161, §103; 42 U.S.C. § 16613(b(f)(4), Congress required NASA to report cost and schedule baselines—benchmarks against which changes can be measured— for all NASA

programs and projects with estimated life cycle costs of at least $250 million that have been approved to proceed to implementation.

[9] GAO, *Tactical Aircraft: F-22A Modernization Program Faces Cost, Technical, and Sustainment Risks*, GAO-12-447, (Washington D.C.: May 2, 2012).

[10] NASA, *Columbia Accident Investigation Board*, Report Volume I, August 2003.

[11] NASA, Statement by Charles Bolden, NASA Administrator, February 1, 2010, NASA Budget Press Conference.

In: Human Space Exploration
Editor: Darrel Gibbs

ISBN: 978-1-63482-172-8


*Chapter 2*

# SPACE LAUNCH SYSTEM: RESOURCES NEED TO BE MATCHED TO REQUIREMENTS TO DECREASE RISK AND SUPPORT LONG TERM AFFORDABILITY*

*United States Government Accountability Office*

## WHY GAO DID THIS STUDY

SLS is NASA's first exploration-class heavy lift launch vehicle in over 40 years. Predecessor programs, such as Constellation, were canceled in the face of acquisition problems and funding shortfalls. NASA estimates it could spend almost $12 billion developing the first of three SLS vehicle variants and associated ground systems through initial launch in late 2017 and potentially billions more to develop increasingly capable vehicles. Ensuring that this program is affordable and sustainable for the long term is a key goal of the 2013 National Space Transportation Policy.

GAO was asked to evaluate SLS program challenges. This report examines (1) the SLS program's progress toward and risks for its first test flight in 2017 and (2) the extent to which the SLS program has plans in place to achieve its long-term goals and promote affordability. To do this, GAO reviewed relevant design, development, cost, and schedule documents;

* This is an edited, reformatted and augmented version of a United States Government Accountability Office publication, No. GAO-14-631, dated July 2014.

interviewed program officials; and evaluated SLS program actions using acquisition and cost estimating best practices.

## WHAT GAO RECOMMENDS

Among other actions to reduce risk and allow for continued assessment of SLS progress and affordability, GAO recommends that NASA develop an executable business case for SLS that matches resources to requirements, and provide to the Congress an assessment of the SLS elements that could be competitively procured for future SLS variants before finalizing acquisition plans for those variants. NASA concurred with GAO's recommendations.

## WHAT GAO FOUND

The Space Launch System (SLS) program is making solid progress on the SLS design. However, the National Aeronautics and Space Administration (NASA) has not developed an executable business case based on matching the program's cost and schedule resources with the requirement to develop the vehicle and conduct the first flight test in December 2017 at the required confidence level of 70 percent. NASA uses a calculation referred to as the "joint cost and schedule confidence level" to estimate the probable success of a program meeting its cost and schedule targets. NASA policy usually requires a 70 percent confidence level for a program to proceed with final design and fabrication. GAO's work on best practices has shown that programs that do not establish an executable business case that matches requirements—or customer needs—to resources, such as schedule and funding—are at increased risk of cost and schedule growth. The program is satisfying many of NASA's metrics that measure progress against design goals, such as requirements for design maturity. According to the program's risk analysis, however, the agency's current funding plan for SLS may be $400 million short of what the program needs to launch by 2017. Furthermore, the development schedule of the core stage— which drives the SLS schedule—is compressed to meet the 2017 launch date. NASA also faces challenges integrating existing hardware that was not originally designed to fly on SLS. For example, SLS is using solid rocket boosters from the Constellation program, but integrating a new nonasbestos insulating material into the booster design has proven difficult and required changes to the booster manufacturing processes.

The SLS program has not yet defined specific mission requirements beyond the second flight test in 2021 or defined specific plans for achieving long-term goals, but the program has opportunities to promote affordability moving forward. NASA plans to incrementally develop more capable SLS launch vehicles to satisfy longterm goals, but future missions have not been determined, which will directly affect the program's future development path and flight schedule. Mission selection will likely determine which element the program decides to develop next, as the program can afford to develop only one element at a time. The magnitude of these development efforts could be significant but is currently unknown as the program has not developed complete life-cycle cost estimates for the initial or future SLS launch vehicles. In May 2014, GAO recommended that NASA address this issue, and NASA partially concurred, citing that actions taken to structure the programs and track costs met the intent of the recommendations. However, GAO believes NASA's responses do not fully address the concerns about the program's cost estimates. There are opportunities, however, to improve long-term affordability through competition once the development path has been determined and NASA can finalize its acquisition approach. For example, the program plans to compete the procurement of one element; however, the agency has not finalized assessments of options for competitively procuring other future elements. Such assessments could better position NASA to sustain competition, control costs, and better inform Congress about the long-term affordability of the program. GAO's body of work on contracting has shown that competition in contracting is a key factor in controlling cost.

## Abbreviations

| | |
|---|---|
| CDR | critical design review |
| EM-1 | Exploration Mission 1 |
| EM-2 | Exploration Mission 2 |
| EVM | earned value management |
| ICPS | Interim Cryogenic Propulsion Stage |
| JCL | Join Cost and Schedule Confidence Level |
| KDP | key decision point |
| MDR | mission definition review |
| mt | metric ton |

| | |
|---|---|
| NASA | National Aeronautics and Space Administration |
| NFS | National Aeronautics and Space Administration Federal Acquisition Regulation Supplement |
| Orion | Orion Multi-Purpose Crew Vehicle |
| PDR | preliminary design review |
| SDR | system definition review |
| SIR | system integration review |
| SLS | Space Launch System |
| UCA | undefinitized contract action |

* * *

July 23, 2014

The Honorable Tom Coburn, M.D.
Ranking Member
Committee on Homeland Security and Governmental Affairs
United States Senate

Dear Senator Coburn:

The National Aeronautics and Space Administration (NASA) is in the midst of developing its first exploration-class heavy lift launch vehicle in over 40 years—the Space Launch System (SLS). The SLS will expand NASA's exploration capability to include crewed flights beyond Earth's orbit. NASA's attempts over the past two decades to develop a successor to the Space Shuttle have been unsuccessful. Prior development programs, the most recent being the Constellation program, were canceled in the face of acquisition problems and funding-related issues. The nearly $12 billion in funding that NASA estimates it could spend developing SLS and associated ground systems through its first launch in late 2017 represents not only a significant portion of NASA's planned budget for major projects during that period but also a significant portion of governmentwide launch-related research and development funding. Developing an exploration program that will be affordable and sustainable for the long term is a key goal of the 2013 National Space Transportation Policy.[1]

GAO has designated NASA's management of acquisitions as a high-risk area for more than two decades in view of persistent cost growth and schedule slippage in the majority of its major projects.[2] While the agency has made progress in recent years in reducing risk on smaller-scale, less complex projects, demonstrating that this progress can be translated to larger, more complex projects, such as SLS, is important. In light of these issues and the long-term significance of the program, you requested that we review the challenges facing NASA's SLS program. This report examines (1) NASA's progress toward the first SLS test flight in 2017 and any acquisition risks associated with its plans and (2) the extent to which the SLS program has plans in place to achieve its long-term goals and promote affordability.

In order to assess NASA's progress to conduct its first flight in 2017, we interviewed and obtained briefings and relevant documents from NASA and contractor officials. We identified and evaluated technical and programmatic issues associated with each major SLS subsystem, by reviewing development plans and discussing relevant issues with agency officials. We also compared planned to actual progress in maturing system designs. We assessed NASA's risk mitigation plans to gauge progress in addressing technical issues and to evaluate the potential impact to scheduled events such as delivery and flight dates. We also reviewed other technical and programmatic indicators and progress made addressing required actions from programmatic reviews at both the subsystem and vehicle level. We evaluated the program's progress toward the 2017 launch date by comparing actual cost, schedule, and performance to current program baselines. To assess long-term affordability, we discussed long-term development plans in support of future missions with agency officials. For purposes of assessing the cost estimate, we reviewed NASA preliminary cost estimates for the SLS, Orion crew capsule that will launch atop SLS, and associated ground systems programs and information related to the baseline cost estimates to determine the scope of the estimates. We assessed the estimates' scope against best practices criteria outlined in GAO's cost estimating guidebook as well as NASA's own requirements and guidance.[3] Furthermore, we evaluated the program's development and acquisition plans to compete future variants of the SLS by reviewing contract information, including any follow-on contract options, and discussing supplier availability with agency officials. For more information on our scope and methodology, see appendix I.

We conducted this performance audit from June 2013 to July 2014 in accordance with generally accepted government auditing standards.

Those standards require that we plan and perform the audit to obtain sufficient, appropriate evidence to provide a reasonable basis for our findings and conclusions based on our audit objectives. We believe that the evidence obtained provides a reasonable basis for our findings and conclusions based on our audit objectives.

## BACKGROUND

The National Aeronautics and Space Administration Authorization Act of 2010 directed NASA to, among another things, develop a Space Launch System as a follow-on to the Space Shuttle and as a key component in expanding human presence beyond low-Earth orbit. To that end, NASA plans to incrementally develop three progressively more capable SLS launch vehicles—70-, 105-, and 130-metric ton (mt) variants. When complete, the 130-mt vehicle is expected to have more launch capability than the Saturn V vehicle, which was used for Apollo missions, and be significantly more capable than any recent or current launch vehicle. The act also directed NASA to prioritize the core elements with the goal of operational capability for the core elements not later than December 2016.[4] NASA subsequently negotiated an extension of that date, to December 2017, based on the agency's assessment of the tasks associated with developing the new launch vehicle.

In 2011, NASA formally established the SLS program. To fulfill the direction of the 2010 act, the agency plans to develop the three SLS launch vehicle capabilities, complemented by the Orion Multi-Purpose Crew Vehicle (Orion) to transport humans and cargo into space. The first version of the SLS that NASA is developing is a 70-mt launch vehicle known as Block I. NASA expects to conduct two test flights of the Block I vehicle—the first in 2017 and the second in 2021. The vehicle is scheduled to fly an uncrewed Orion some 70,000 kilometers beyond the moon during the first test flight, known as Exploration Mission-1 (EM-1), and to fly a second mission known as Exploration Mission-2 (EM-2) beyond the moon to further test performance with a crewed Orion vehicle. After 2021, NASA intends to build 105- and 130-mt launch vehicles, known respectively as Block IA/B and Block II, which it expects to use as the backbone of manned spaceflight for decades. [5] The agency plans for these vehicles to carry larger cargo and travel farther into space, but it has not yet selected specific missions for the increased capabilities to be provided by Block IA/B and Block II. NASA anticipates using the Block IA/B vehicles for destinations such as near-Earth asteroids and LaGrange points and the Block II vehicles for eventual Mars missions.[6]

## NASA's Acquisition Life Cycle

NASA plans to evolve the 70-mt SLS design following the agency's life cycle acquisition process for flight systems. That process is defined by two broad phases—formulation and implementation—and several key decision points.[7] These broad phases are then further divided into more discrete pieces with different purposes: pre-phase A through phase F. See figure 1 for a depiction of NASA's life cycle for flight systems.

During formulation, programs develop and define requirements and establish cost and schedule baselines and an acquisition strategy. Formulation concludes with a preliminary design review (PDR) and project confirmation review where cost and schedule baselines are confirmed and documented in the agency baseline commitment.[8] The PDR evaluates the completeness and consistency of the planning, technical, and cost and schedule baselines developed during formulation. It assesses compliance of the preliminary design with applicable requirements, and determines if the project is sufficiently mature to begin the final design and fabrication phase. The agency baseline commitment establishes and documents an integrated set of requirements, cost, schedule, technical content, and an agreed-to joint cost and schedule confidence level (JCL) that forms the basis for NASA's commitment with the Office of Management and Budget and the Congress.[9] These baselines are informed by the JCL process. The JCL is a probabilistic analysis that provides assurance to stakeholders that programs will meet cost and schedule targets. In general, programs' cost and schedule baselines are based on a 70 percent confidence level, unless the decision authority approves a different confidence level with appropriate justification and documentation.[10] This is the point on the joint cost and schedule probability distribution where there is a 70 percent probability that the project will be completed at or lower than the estimated amount and at or before the projected schedule. Program progress can subsequently be measured against the baseline commitments.[11] NASA's policy requires that a JCL be developed at a program's confirmation review.[12] According to agency officials, NASA is currently establishing the JCL for SLS and the program will move into the implementation phase when the confirmation review is complete.

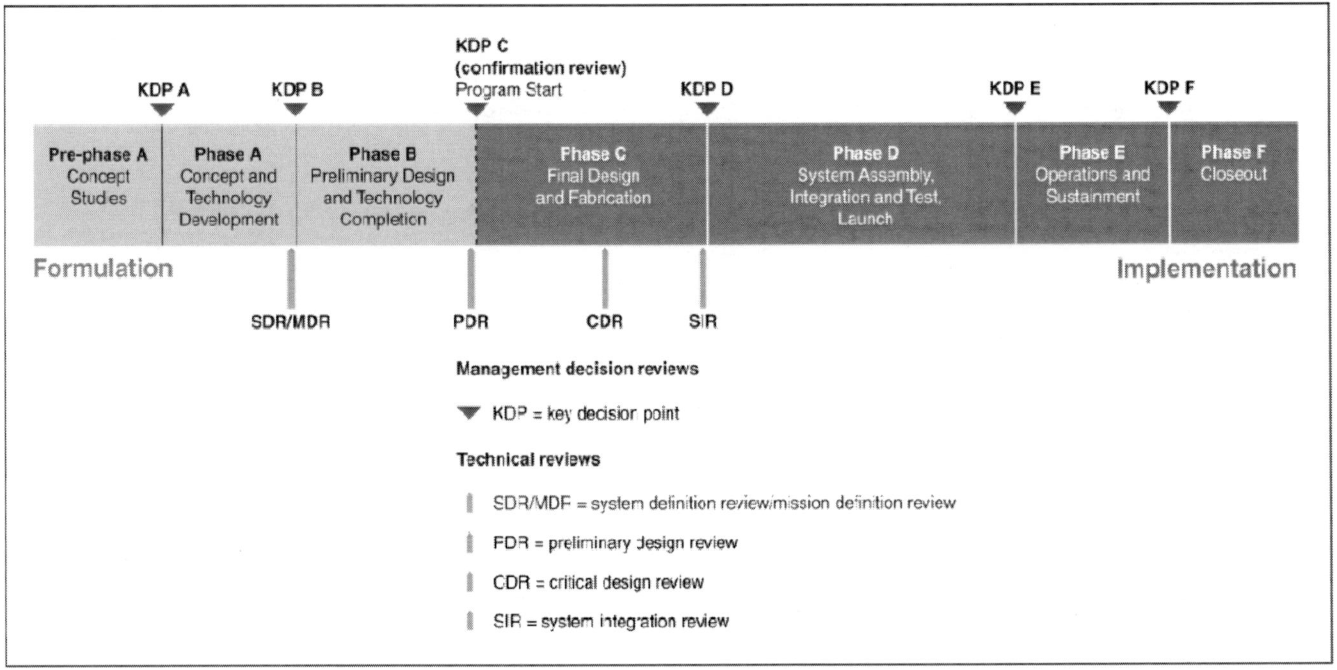


Figure 1. NASA's Life Cycle for Flight Systems.

## Acquisition Best Practices

GAO's best practices for systems acquisition state that evolutionary acquisition is a commercial best practice.[13] In evolutionary acquisition, a program uses mature technologies and capabilities that fit within existing resources to field initial capabilities sooner and gradually develops progressively more capable increments until it achieves its ultimate requirements. This approach is considered a best practice because it allows commercial companies and the government to develop and produce more sophisticated products faster and less expensively.

GAO's work on best practices has also shown that success in development efforts such as SLS depends on establishing an executable business case based on matching requirements and resources before committing to a new product development effort. In its simplest form, a business case requires a balance between the concept selected to satisfy customer needs and the resources—technologies, design knowledge, funding, time, and management capacity—needed to transform the concept into a product. At the heart of a business case is a knowledge-based approach that requires that managers demonstrate high levels of knowledge as the program proceeds from technology development to system development and, finally, production. Ideally, with such an approach, key technologies are demonstrated before development begins, the design is stabilized before prototypes are built or production begins, and testing is used to validate product maturity at each level. At each decision point, the balance among time, money, and capacity is confirmed. In essence, knowledge supplants risk over time. Having adequate knowledge about requirements and resources is particularly important for a program like SLS because human spaceflight development projects are complex, difficult, and costly.

We have found that within NASA's acquisition framework, the PDR and corresponding confirmation review are the point at which development projects should have a sound business case in hand.[14] NASA's Systems Engineering Policy states that this review demonstrates that the preliminary design meets all system requirements with acceptable risk and within the cost and schedule constraints. After a project is confirmed following PDR, it begins implementation during which time senior NASA officials periodically review the programmatic and technical status of projects and verify the project's readiness to proceed forward. For example, a critical design review (CDR) before the project's subsystems are integrated evaluates the integrity of the project design and its ability to meet mission requirements, with appropriate margins and acceptable risk, within defined project constraints, including available resources. In short, the CDR determines if the design is appropriately mature to support proceeding with the final design and fabrication phase.[15] Our past work on product development best practices has found that programs having at least 90 percent of engineering drawings releasable by the critical design review lower their risk of subsequent cost growth and schedule delays, and guidance in NASA's *Systems Engineering Handbook* mirrors this metric.[16] NASA also tracks both mass margin and resolution of requests for action, which are technical, safety, and programmatic comments from independent reviewers, as additional measures of design maturity.[17] In addition, experts in the space community have identified other metrics that can be useful to assess the design stability of unique space systems.[18] These include the program's level of funding reserves and schedule margin at various points in the development life cycle as well as the percent of verification and validation plans completed at both PDR and CDR.

## Elements of SLS

In accordance with direction contained in the NASA Authorization Act of 2010, NASA's acquisition approach for building the initial variant of the SLS is predicated on the use of legacy systems, designs, and contracts from the Space Shuttle and Constellation programs. Figure 2 provides details about the heritage of each SLS hardware element and its source.

NASA plans to use 16 RS-25 engines remaining from the Space Shuttle program to provide power for up to four flights of the SLS. Similarly, the agency is procuring five-segment boosters that were developed under the Constellation program to provide thrust during the initial minutes of SLS flight. NASA is also procuring a cryogenic rocket stage used on United Launch Alliance's Delta IV launch vehicle and modifying it to operate as the Interim Cryogenic Propulsion Stage (ICPS) to provide in-space power for SLS. Finally, the design for the new core stage, which functions as the SLS's fuel tank and structural backbone, is derived from the Shuttle's external tank and Ares I upper stage.

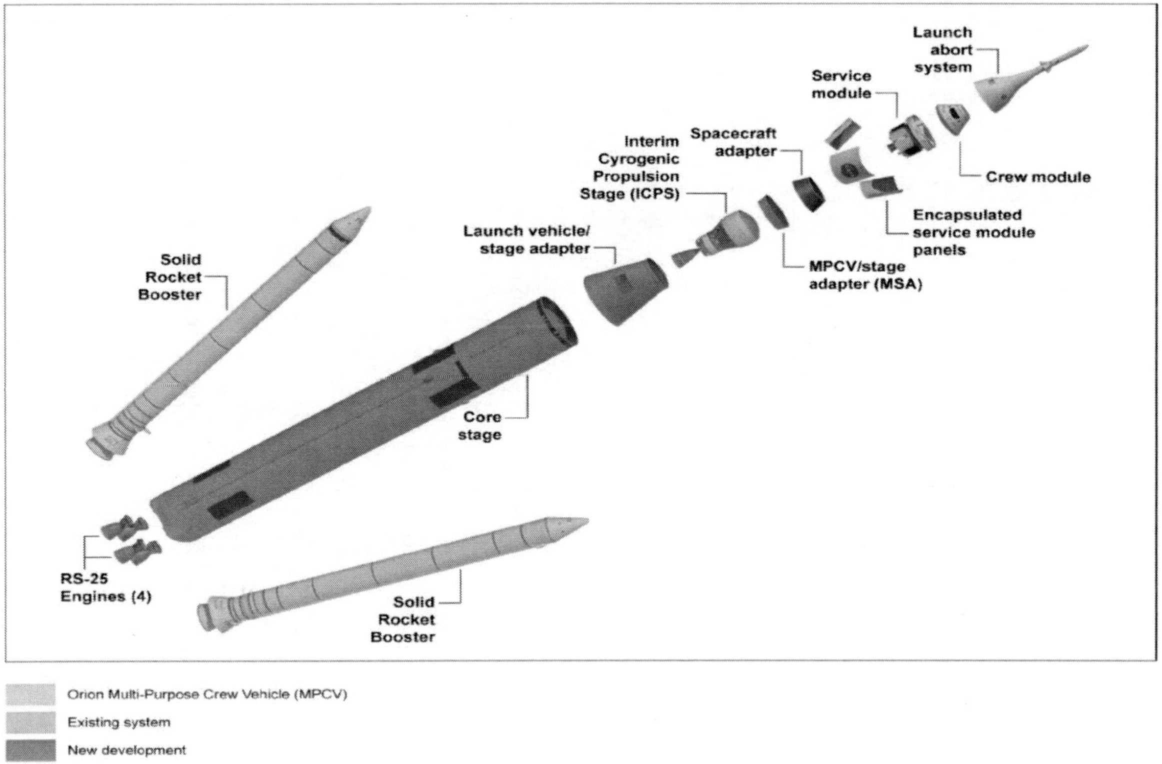


Figure 2. SLS and Orion Hardware Elements.

# PROGRAM'S ABILITY TO MEET SCHEDULE FOR 2017 TEST FLIGHT AT RISK

While the SLS program is satisfying many of NASA's metrics that measure progress against overall design goals, NASA has not established an executable business case that matches the SLS program's cost and schedule resources with the requirement to develop the SLS and launch the first flight test in December 2017 at the required confidence level of 70 percent.[19] Matching resources to requirements is considered a best practice for establishing a successful acquisition program.[20] Reduced confidence levels increase the risk that the program will miss its launch date or overrun its cost baseline. The program is on target to achieve goals for design drawing release and mass and expects to meet necessary documentation requirements for its critical design review. However, the development schedule of the core stage, which is driving the overall program schedule, is aggressive and substantial amounts of schedule that the program reserved to resolve unanticipated issues is already threatened. NASA has focused on integration issues, and the program is on plan for closing design trades, environmental studies, and design interface documents that are intended to reduce this risk. However, the modification and integration of heritage elements may be a challenge—one which NASA often underestimates. In addition, according to the program's risk analysis, the agency's current funding plan for SLS may be $400 million short of what the program needs to launch by 2017. This funding discrepancy was, in part, responsible for the significant delay in finalizing the contracts for element development, which increased cost and schedule risk to the government as well as reduced visibility into contractor performance.

## NASA Has Not Matched Resources and Requirements for 2017 Initial Flight Test

NASA has not established an executable business case based on matching the SLS program's cost and schedule resources with the requirement to develop the SLS and launch the first flight test in December 2017 at the required confidence level of 70 percent. NASA delayed the SLS key decision point C decision from October 2013 to at least July 2014, as the agency considered future plans for the program. If the agency determines the current funding plan for SLS is insufficient to match requirements to resources for the

December 2017 flight test at the 70 percent confidence level, the agency's options for matching resources to requirements are largely limited to increasing program funding, delaying the schedule or accepting a reduced confidence level for the initial flight test. While the program's decision authority is allowed in some instances to approve a confidence level of less than 70 percent, doing so increases the likelihood that the program will miss the launch date or overrun the current cost estimate. Should cost growth or schedule delay occur, it could place other programs' funding at risk if NASA chooses to take planned funding from those programs in order to maintain the SLS schedule. Although cost and schedule growth can occur on any project, increases associated with NASA's most costly and complex missions— such as SLS, which makes up about 9 percent of NASA's annual budget—can have dramatic effects on the availability of funding for NASA's portfolio of major projects.

Funding levels have impaired the SLS program's ability to match requirements to resources since its inception. As illustrated in figure 3, NASA has requested relatively consistent amounts of funding for SLS each year. According to agency officials, the program has taken steps to live within that flat funding profile, including streamlining program office operations and asking each contractor to identify efficiencies in their production processes. Even so, according to the program's own analysis, going into the confirmation review, SLS's top risk was that the current planned budget through 2017 would be insufficient to allow the SLS as designed to meet the EM-1 flight date. The SLS program office calculated the risk associated with insufficient funding through 2017 as having a 90 percent likelihood of occurrence; furthermore, it indicated the insufficient budget could push the planned December 2017 launch date out 6 months and add some $400 million to the overall cost of SLS development. Program officials stated that the potential cost impacts of this risk were considerably higher in the past, but that they were able to reduce the impact due to receiving more funding than requested in fiscal years 2013 and 2014.

In May 2014, the House Committee on Appropriations report—which accompanied the Commerce, Justice, Science, and Related Agencies Appropriations Bill, 2015—"expressed frustration with NASA's practices of requesting arbitrarily reduced funding levels for SLS and insisting that the program manage to an inefficient flat-line budget profile." The Committee noted that this practice has "detrimental results" and is likely to lead to a "launch delay for EM-1 and deferral of long-lead work needed for the timely achievement of EM-2 and other future flights." In light of these and other issues, the Committee has chosen to recommend maintaining SLS vehicle development funding at the fiscal year 2014 enacted level of $1.6 billion.

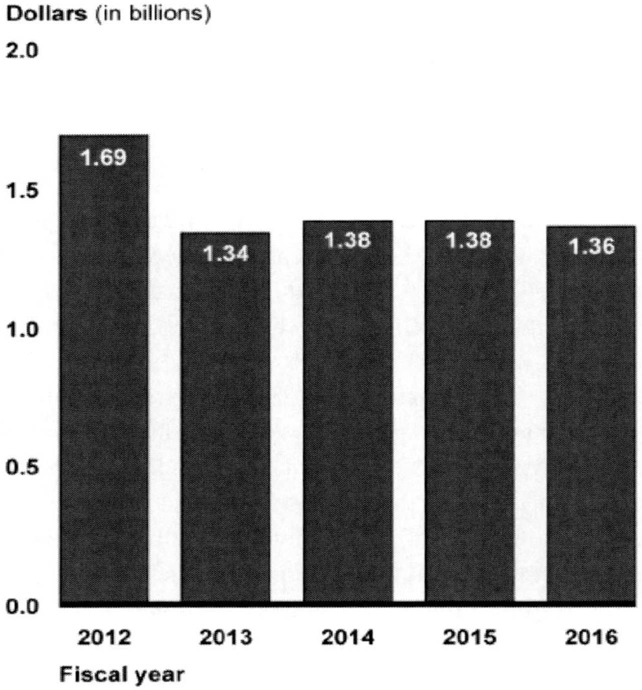


Source: GAO analysis of NASA budget data. | GAO-14-631.
Note: Fiscal year 2016 funding estimate is notional.

Figure 3. SLS Funding Requested, Fiscal Years 2012-2016.

Best practices for acquisition programs indicate that matching resources to requirements and rationally balancing cost, schedule, and performance is a key step in establishing a successful acquisition program.[21] We have also found that NASA's previous attempts to develop new transportation systems, such as the Constellation program, have failed in part because they were focused on maturing designs without adequate funding to support those efforts.[22] Our work has also shown that developing a sound business case before committing resources to development—which includes firm requirements, mature technologies, a knowledge-based acquisition strategy, a realistic cost estimate, and sufficient funding and time—would help mitigate the risks inherent in NASA's programs.[23]

## SLS Program Is Meeting Some Design Goals for 2017 Test Flight but Challenges Could Threaten Launch Date

### *Design Goals*

NASA's metrics indicate that the program is on track to meet many of its design goals for demonstrating the initial capability of SLS. Based on our review of top level design metrics that the program is tracking, the program is currently meeting its plan for design drawing release for the core stage, where most of the new design work is occurring. As of May 2014 the program had released 82 percent of the core stage drawings and expects to have 95 percent released by the core stage CDR in July 2014. The program as a whole expects to release 95 percent of its design drawings by the program level CDR in March 2015, which exceeds the best practice metric for design drawing release at CDR.[24] Because the CDR is the time in a project's life cycle when the integrity of the project design and its ability to meet mission requirements are assessed, it is important that a project's design is stable enough to warrant continuation with design and fabrication, which is evidenced by release of 90 percent of design drawings at CDR. A stable design allows projects to "freeze" the design and minimize changes prior to beginning the fabrication of hardware. It also helps to avoid re-engineering and rework efforts due to design changes that can be costly to the project in terms of time and funding.

The program is also meeting or close to meeting several design goals related to SLS mass margin and program documentation. As of May 2014, SLS had sufficient mass margin to meet its design goals for EM-1; however, the program has less mass margin for some EM-2 test flight mission options in 2021. If mass becomes an issue leading into EM-2, the vehicle may require design changes that could lead to cost and schedule growth. While the program is behind on its own schedule for addressing the requests for action from the preliminary design review, the program's current plan indicates that all requests will be addressed prior to the CDR scheduled for 2015. In addition to these overall design goals, the program's verification and validation plan, which aids in assessing whether a chosen development solution is capable of meeting its intended purpose and if the system is being developed according to agency requirements, was baselined and approved at PDR with one exception that the program expects to resolve in 2014. With regards to finalizing interface control documents, which define requirements for how systems will connect and interact with one another, the program is 4 percent behind its planned schedule for determining interface requirements by CDR. According to NASA, these documents should help reduce integration issues, such as those between the heritage hardware elements and new core stage, by specifying interaction between the heritage elements and core stage, as well as provide a set of interface definitions for when new elements are being designed.

### *Aggressive Schedule*

Although the program is making progress against its design metrics, the core stage development schedule is aggressive and already threatened, and any delays could impair SLS readiness for first flight in 2017. The core stage development drives the SLS schedule because it represents the critical path of activities that must be completed to maintain the program's schedule as a whole. Based on our review of NASA documentation and discussions with program officials, it is clear that the core stage's major program milestones and developmental activities are tightly spaced. For example, the program is allowing 18 months from the point at which it evaluates the design at PDR in January 2013 to the point at which it expects the design to be fully mature at the core stage CDR in July 2014. By comparison, the Ares I project within the Constellation program planned for about 2 years between PDR and CDR. NASA officials indicated that they compressed the core stage development schedule in order to meet the SLS 2017 launch date. The schedule allows for 7 months of schedule reserve after CDR and before the core stage ships to Kennedy Space Center for integration into the SLS, which NASA believes will allow time to address problems it could encounter during development. While allowing 7 months of schedule reserve generally complies with NASA guidance for space flight projects, it may prove insufficient in this instance because the program has already identified a number of threats to the schedule.[25]

The SLS program is tracking threats to the core stage schedule that could take up as much as 70 percent of the 7 months of reserve. Many of these threats are associated with the schedule for acquiring liquid oxygen feeder lines that provide liquid oxygen from the fuel tanks within the core stage to the core stage engine. Agency officials indicated the program is encountering difficulties acquiring feeder lines from available suppliers to meet the core stage development schedule because the feeder lines are larger and will be used in more stressful environments than the lines that suppliers have manufactured in the past. The feeder lines must still be designed, built, and tested for operation in the SLS environment before they can be delivered for integration into the system. While these challenges are not overly complex from a technical viewpoint, resolving such threats to the schedule is critical because the element is in early development phases and still has several significant milestones and developmental activities ahead. For example, the core stage has CDR currently scheduled for July 2014 that will assess the program's design maturity. Further, qualification testing to determine if system components meet requirements is currently scheduled to begin in January 2015. As we have found in the past, the integration and test phase of development is generally where issues that require unexpected time and resources to address are discovered.[26]

### *Heritage Hardware*

The SLS program could experience additional schedule pressure if unanticipated challenges associated with using heritage hardware occur when integrating it into the launch vehicle's operational environment and modifying manufacturing process to incorporate new materials. The use of heritage hardware—legacy engine, booster, and propulsion systems— was prescribed in the NASA Authorization Act of 2010, but the hardware was not originally designed for SLS. Therefore, the SLS program must ensure each heritage hardware element meets SLS performance requirements and current design standards prior to the 2017 test flight. Although the heritage hardware challenges have yet to affect the SLS schedule, each heritage hardware element shares the common issue of operating in the SLS environment that is likely to be more stressful than that of its original launch vehicle as well as unique integration issues particular to that element, which must be resolved prior to SLS first flight in 2017. For example, according to agency officials the engines from the Space Shuttle require additional heat shielding because of the increased temperatures they will experience in the SLS environment, and the avionics within the solid rocket boosters from the Constellation program are likely to require additional cushioning to protect them from increased vibrations. Until the core stage is demonstrated, however, the SLS operating environment can be defined only through analytical predictions. Further, eliminating asbestos as a key insulating material within the solid rocket boosters on the SLS has required changes to the booster manufacturing processes to meet safety requirements.

According to agency officials, the challenges they face are typical of those expected when heritage hardware is integrated into a new system. They also noted that addressing these challenges involves less time, money, and overall effort than developing new hardware and that each heritage hardware element has schedule reserves to address anticipated integration challenges as well as unknown issues that might arise. Nevertheless, the engineering effort required to address them is significant and, as found by both GAO's and the NASA Inspector General's prior work, the complexity associated with required modifications and problems with availability of components used on projects is often underestimated.[27]

## Unclear Scope and Funding Uncertainties Increased Risk by Delaying Contract Definitization

Uncertainty regarding resources increased risk to the program because contracts were not finalized and complete contractor performance data was not provided. According to agency officials, the program was delayed in definitizing contracts, in part, due to resource uncertainty. As a result, the SLS program contractors have worked for extended periods of time without contract definitization—meaning no final agreement on the terms and conditions of their contracts has been negotiated with the government—which put the government at risk of increased costs and limited the program's ability to monitor contractor progress. Contractors have performed, and are performing, work under undefinitized contract actions (UCA) until their respective contracts are definitized and cost and schedule baselines are set. Contract actions such as these authorize contractors to begin work before reaching a final agreement with the government on contract terms and conditions. The government is thus in a weaker position to control costs.

The NASA supplement to the Federal Acquisition Regulation (NFS) provides that the NASA goal is to definitize UCAs within 180 days of issuance, or approximately 6 months.[28] Officials stated that the agency's ability to definitize the element contracts was impacted by the need to use existing contracts, as directed by the NASA Authorization Act of 2010, and to complete work necessary to define requirements and modify those contracts in order to accommodate the SLS effort's scope and content.

The NFS allows for such use of UCAs on an exception basis and also notes that, when UCAs are used, the agency's liabilities and commitments must be minimized.[29] As figure 4 illustrates, the booster and main engine element contractors performed development work under UCAs for more than a year and the core stage and ICPS contractor continued those elements' development under UCAs for 2 years or more.

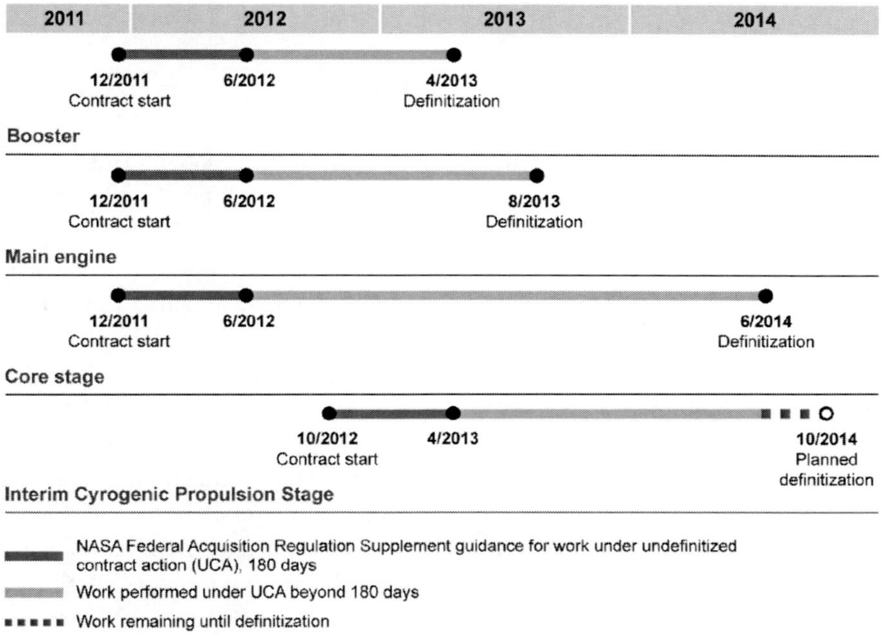


Source: GAO analysis of NASA budget data. | GAO-14-631.

Figure 4. Months Each SLS Element Spent under Undefinitized Contract Actions.

NASA allowed high-value modifications to the SLS contracts to remain undefinitized for extended periods—in one instance a contract remained undefinitized for 30 months. Because lack of agreement on contract terms prolonged NASA's timeframes for finalizing SLS contracts, the establishment of contractor cost and schedule baselines necessary to monitor performance was delayed. Specifically, in most cases, the SLS program did not receive complete earned value management (EVM) data derived from approved baselines on these SLS contracts. Earned value, or the planned cost of completed work and work in progress, can provide accurate assessments of project progress, produce early warning signs of impending schedule delays and cost overruns, and provide unbiased estimates of anticipated costs at completion. The use of EVM, which integrates the project scope of work with cost, schedule, and performance elements for optimum project planning and control, is advocated by both GAO's best practices for cost estimating and NASA's own guidance.[30] Without having a contracted baseline against which to measure the earned value, the program is missing a useful indicator of true program status—in the case of SLS, the program's performance toward its 2017 launch date.

The program began receiving EVM data from the booster contractor in 2013, and began receiving it from the main engine contractor in 2014, but has yet to receive EVM data from the core stage or ICPS contractors because, according to program officials, the core stage contract was just definitized and the ICPS contract has yet to be definitized and baselines have yet to be established. We have previously found that UCAs transfer cost and performance risk from the contractor to the government. [31] Our earned value body of work indicates that an integrated baseline review can help allay that risk by providing the government with assurance that the performance baseline reflects all requirements and that resources are adequate to complete the work.[32] Program officials stated that they planned to hold an integrated baseline review for each element, following contract definitization, in order to begin receiving EVM data from the element contractors. In lieu of EVM, according to program officials, the contractors have been providing cost and schedule data to NASA derived from their internal baselines. Until the program and the contractor reach final agreement on contract terms, however, this data may not reflect whether the contractor is accomplishing the work planned within the agency's allocated cost and schedule. Until baselines are set and an integrated baseline review is held, the cost and schedule data can only be used as indicators of progress. Furthermore, such data can be misleading as it lacks a comparison between work performed versus planned.

## SLS PROGRAM HAS CRITICAL GAPS IN KNOWLEDGE NEEDED TO ASSESS LONG-TERM AFFORDABILITY, BUT OPPORTUNITIES EXIST TO PROMOTE AFFORDABILITY

NASA has yet to make mission decisions beyond EM-2 for the SLS program and has not produced a complete life-cycle cost estimate for any of the three planned variants; however, competition opportunities exist for future development work that may promote long-term affordability. Although the agency has identified several possible destinations, it has not decided upon specific missions for the SLS program beyond EM-1 and EM-2, which will directly affect the program's future development path and flight schedule. Mission selection will likely determine the order of future development efforts, as the program can only afford to develop one upgraded element at a time. Each development effort may be large enough to constitute a separate project, but the full magnitude of these development efforts is unknown as the program has not established cost estimates for any of the three SLS variants, including the 70-mt vehicle, beyond 2017. Although the program costs are as yet unknown, there are opportunities to improve long-term affordability through competition once the element upgrade development path has been determined. While the main engine contractor for future SLS vehicles is likely to remain the same, the program plans to compete the acquisition of the advanced boosters. Additionally, the program could compete the acquisition of several other elements in order to potentially reduce costs.

## The SLS Program's Long-Term Mission Plans and Requirements Are Unknown

NASA has not yet defined specific mission requirements for any variant of the SLS. The two currently scheduled flights, EM-1 and EM-2, are developmental test flights designed to demonstrate and test the capabilities of the 70-mt launch vehicle and the capability of the core stage in particular. Office of Management and Budget guidance indicates that agencies should develop long range objectives, supported by detailed budgets and plans that identify the agency's performance gaps and the resources needed to close them.[33] According to agency officials, beyond the two scheduled test flights, no future mission destinations have been determined. In the absence of specific mission requirements, officials indicated the SLS program is developing the variants based on top-level requirements derived from NASA's Design Reference Architectures that lay out the technical and scientific framework for conducting missions in line with the agency's strategic plan. NASA's 2014 strategic plan, for example, identifies sending humans to Mars as one of the agency's long-term goals; in turn, the agency's Mars Design Reference Architecture indicates that multiple missions using a vehicle with a lift capability of about 130 mt will be necessary to support that goal.

Although NASA is considering long-term Mars missions and asteroid redirect missions, NASA has not yet finalized plans for the next step in evolving the SLS and risks investing limited available resources in systems and designs that are not yet needed. The agency intends to use the same core stage design for all variants of the SLS, but it will be faced with replacing all the other major hardware elements used on the 70-mt vehicle—that is, the upper stage, boosters, and engines—so that the 105- and 130-mt vehicles can perform missions that support the agency's strategic goals of carrying heavier payloads and traveling deeper into space. To that end, NASA is working on analyses and concept designs for a new upper stage and for advanced boosters. Because mission requirements for the 105-mt variant are not yet defined, however, NASA faces the possibility that it will spend time and money developing systems and designs that may not be needed for the specific mission that will be defined. For example, according to program officials and as illustrated in figure 5, in order to have a 105-mt capability ready to launch soon after 2021, the agency must make a decision not later than 2016 about whether it will first pursue development of the upper stage or advanced boosters as the next step in increasing SLS capability beyond the 70-mt variant.

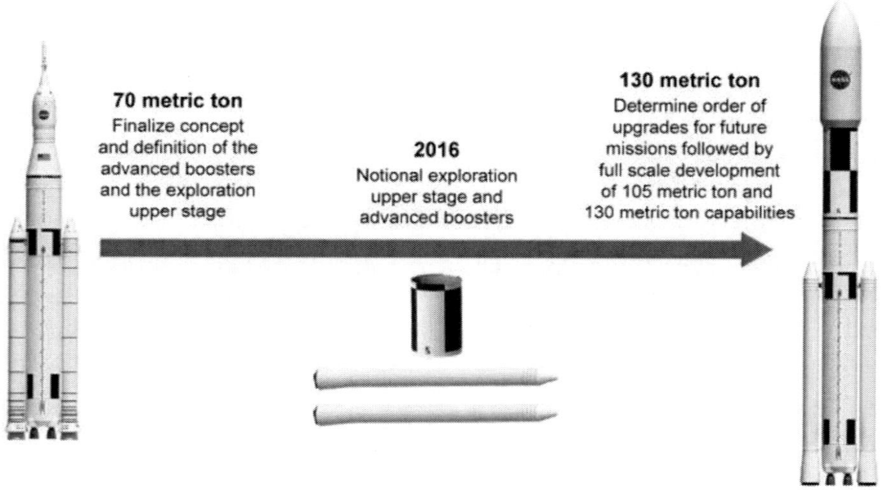


Source: GAO presentation of NASA data. | GAO-14-631.

Figure 5. SLS 2016 Vehicle Development Decision.

Program officials stated that the agency does not have resources to conduct more than one development effort at a time and that it is difficult to determine which system to develop first because each improves SLS lift capability in different ways. A new upper stage would provide more capability beyond Earth orbit and better support missions that require more in-space propulsion, such as missions to a near-Earth asteroid or other distant locations. Advanced boosters would provide more capability to low-Earth orbit and better support missions that require the SLS to place a larger payload in orbit around the Earth. No matter which effort NASA pursues, in order for the agency to begin full-scale development after 2016, it is confronted with requesting related funding in the year before it plans to begin work. For example, if NASA intends to initiate full-scale upper stage development in 2017, including the related funding request in its fiscal year 2016 budget submission would be prudent. The sooner NASA makes a decision about which development to pursue first, the sooner it will be able to focus its limited resources on the chosen effort. Conversely, the longer the agency delays the decision, the more likely it will expend those resources on a design that might not yet be needed.

Regardless of when the mission requirements are established and what development NASA first pursues, those future development efforts are likely to be of sufficient cost and scope to constitute programs or projects in and of themselves. According to NASA officials, the advanced booster and new upper stage, for example, will involve distinct development, integration, qualification, and testing efforts. As such, they meet NASA's own criteria for a new project in that they will represent a specific effort with defined requirements, a life-cycle cost estimate, and a beginning and end, and will result in new or revised products, in the form of improved boosters and stages, to address strategic goals. Further, these efforts are likely to far exceed the $250 million life-cycle cost established in NASA policy as the threshold for requiring the NASA Associate Administrator approval of cost and schedule baseline commitments.[34] These projects are subject to oversight by senior agency officials through periodic programmatic reviews that assess technical and programmatic readiness to move forward at various decision points. At this time, future SLS development efforts are not planned to be managed as separate projects, though NASA has indicated that it plans to hold design reviews for these efforts consistent with its acquisition policies and practices. Best practices for system acquisition and cost estimating advocate an evolutionary approach wherein each capability increment has its own cost, schedule, and performance baselines and methods to report progress against these baselines.[35] In addition, we have previously concluded that it is prudent for an agency to manage increasing capabilities of an existing program on par with the level of investments yet to come and in a way that is beneficial for oversight. For example, we have recommended that agencies developing weapon systems in increments consider establishing each increment of increased capability with its own cost and schedule baseline.[36] Establishing cost, schedule, and performance baselines for each increment that has costs exceeding $250 million could provide decision makers with additional insights into the progress and long-term affordability of each increment.

## The Program's Long-Term Affordability Is Unknown Because SLS Cost Estimate Does Not Capture Life Cycle Costs

The long-term affordability of the SLS program is also unknown, as we found in May 2014, because NASA's baseline cost estimate for the program will not provide any information about the longer-term, life cycle costs of developing, manufacturing, and operating the launch vehicle.[37] NASA does not plan for that baseline estimate, which will be established when SLS moves into implementation, to cover program costs after EM-1 or costs to design, develop, build, and produce the 105- or 130-mt variants.

Based on the tenets of widely-accepted best practices for cost estimation as well as NASA's own requirements and guidance regarding life cycle costs, in May 2014 we made recommendations to enhance transparency, assist congressional oversight, and ensure insight into affordability as the agency moves ahead with the SLS, Orion, and related ground support programs. 38 In terms of SLS, we recommended that NASA establish a separate cost and schedule baseline for work required to support SLS Block I EM-2 and report that information via its annual budget submission. As part of that recommendation, we noted that, if NASA decides to fly the SLS Block I 70-mt variant beyond EM-2, the agency should then establish separate life cycle cost and schedule baseline estimates for that variant and report this information via its annual budget submission. Additionally, we recommended that NASA establish life cycle cost and schedule baselines for each upgraded block of the SLS, Orion, and associated ground support. We stated that, if the agency believed it could not do so because missions and flight manifests were unknown, then it should forecast minimum and maximum ranges for the increased capabilities' life cycle costs in its annual budget submission. NASA partially concurred with our recommendations, citing among other reasons that actions already in place such as establishing SLS, Orion, and related ground support as separate programs and a block upgrade approach for SLS— and actions it plans to take to track costs—met the intent of our recommendations.

In our evaluation of NASA's comments, we acknowledged that these actions were a step in the right direction, but noted that NASA's actions do not fully address our concerns. Specifically, we pointed out that establishing cost and schedule at the broader program level was unlikely to provide the detail necessary to monitor the progress of each block against a baseline; it was unclear from NASA's response whether cost commitments the agency plans within the SLS design review process would serve the same purpose as establishing a cost baseline for each respective upgrade; and reporting costs associated with EM-2 and subsequent variants of SLS via the agency's annual budget submission would not provide information about potential costs over the long term because budget requests neither offer all the same information as life-cycle cost estimates nor are necessarily linked to an established baseline that indicates how much NASA expects to invest to develop, operate, and sustain a capability over the long-term. We stated that, as NASA establishes parameters for the additional flights of the first SLS capability and upgraded capabilities, including flight rates, mission destinations, and other requirements, it will be well-poised to move from reporting costs in budget submissions to establishing baseline cost and schedule estimates for each capability and reporting progress against these respective baselines.

## NASA Taking Steps for Long-Term Affordability but Additional Opportunities May Exist

NASA has taken some steps that may promote the long-term affordability of the SLS program as directed in the National Space Transportation Policy.[39] The policy requires that human space flight development efforts such as the SLS program identify and implement measures to enhance affordability. To promote affordability, SLS has funded a series of studies and assessments on advanced boosters by both potential industry partners and universities. According to program officials, these activities—which include improved welding procedures, simulation of material performance, and demonstration of potential propellants—may provide better understanding of the processes, structures, and risk levels of various booster development options, which may in turn reduce the cost or schedule necessary to design and develop an advanced booster. Program officials also indicated that such activities may provide better insight into which contractors may compete for future booster development contracts.

The program, however, could promote longer term affordability by introducing competition in the development and production contracts for other hardware elements. Our body of work on contracting has shown that competition in contracting is a key element for maintaining cost control.[40] We have found that promoting competition increases the potential for acquiring quality goods and services at a lower price and that noncompetitive contracts carry the risk of overspending because, among other reasons, they have been negotiated without the benefit of competition to help establish pricing. NASA has structured other launch programs, such as the Commercial Crew and Launch Services Program, based on the premise that competition will lower costs. In addition, the NASA Authorization Act of 2010 requires that NASA compete sub-elements of the SLS. Furthermore, with certain exceptions, full and open competition in soliciting offers and awarding contracts is generally a requirement of the Competition in Contracting Act of 1984 and the Federal Acquisition Regulation.[41] Given the long-term planned use of SLS, using non-competitive procedures to acquire any portion of the SLS vehicle may lead to higher costs that could be felt for years or decades.

As it replaces hardware elements used on the 70-mt variant with elements necessary to support NASA's anticipated uses of the 105- and 130-mt variants, the SLS program will likely have several opportunities to promote long-term affordability through competing contracts for the new elements. The program plans to use only the core stage and main engines for all flights beyond EM-2. Program briefings indicate that the SLS program optimized the design of the core stage, which will be used with all vehicle variants, to utilize the RS-25 engines remaining from the Space Shuttle program. According to agency officials, a natural consequence of finalizing the core stage design based on the RS-25 is that the only engine-related efforts open to future competition will be RS-25 subsystems and components. The program could be in a favorable position, however, to compete contracts for the exploration upper stage, the upper stage engine, and advanced boosters that it expects to use on the 105- and 130-mt variants. According to SLS program officials, it currently plans to compete contracts for the development and production of new advanced boosters that will be more capable than the current five-segment boosters. The program is considering multiple design solutions including metallic or composite casings and liquid or solid fuel. Based on our review of program documentation and discussions with agency officials, however, additional opportunities exist to compete the procurement of other sub-elements needed for the 105- and 130-mt vehicles. In some cases other programs, both within and outside of NASA, use alternative hardware that serves similar purposes.

- According to program officials, the program plans to procure a new upper stage through the existing core stage contract to replace the ICPS, which the program is using as an interim solution, for all missions following EM-2. They also stated that the program does not plan to compete the upper stage development because its initial award to Boeing in 2007 under the Constellation program was done competitively. Since that award, however, our work indicates that the marketplace for spacecraft development has shifted considerably as new commercial providers have since developed and have launched, or are currently developing, upper stages. These providers may offer viable competition for the new upper stage. For example, in recent years Orbital Sciences Corporation and SpaceX have successfully flown upper stages for NASA's Commercial Cargo program and SpaceX and Boeing will need to human-rate upper stages as part of NASA's Commercial Crew program with flights expected to begin around 2017.

- The SLS program must also procure a new upper stage engine. According to NASA officials, the upper stage engine will be selected based on mission need. Potential competitors include Aerojet Rocketdyne's J-2X, the Evolved Expandable Launch Vehicle's RL10, and the BE-3 engine that Blue Origin is developing for the Commercial Crew program, among other options.

Because, except for the RS-25 engines, NASA's current contracting approach for the SLS program, does not commit the program beyond the hardware needed for EM-2, moving forward the agency will be in the position to take advantage of the evolving launch vehicle market. An updated assessment of the launch vehicle market could better position NASA to sustain competition, control costs, and better inform the Congress about the long-term affordability of the program.

## CONCLUSION

NASA established the SLS program to provide the capability for transporting humans into space, but the agency is unlikely to do so as quickly as it intended. The initial launch date for SLS is just 3 and a half years away. That is a short time given the amount of development, integration, and testing that must occur for the vehicle to fly in 2017. While the technical challenges associated with those efforts appear manageable, the compressed development schedule in conjunction with the agency's relatively flat funding profile for SLS through 2017 place the program at high risk of missing the planned December 2017 launch date for the EM-1 initial test flight.

Beyond the SLS second flight in 2021, the program's path is unclear. NASA is waiting for additional policy direction on future missions, but the agency is approaching a crossroads wherein it is confronted with defining a developmental path forward to the more capable variants of SLS. Without identifying a range of mission possibilities and their required funding, the program is at risk of making uninformed decisions and pursuing development paths that may not make the most efficient use of limited resources in the near term and could negatively impact longer term affordability. Furthermore, without this information the agency's ability to make important decisions about the affordability of the program in the context of the agency's overall budget and competing priorities is limited. Additionally, it is unclear how the agency plans to manage the upgraded capability efforts. Without carefully structuring the approach for future variants, decision makers will lack transparency into costs and will be limited in their ability to assess long-term affordability and progress. This lack of knowledge about program direction and cost is of concern because the success of these programs will be measured by both the capabilities they achieve and their affordability. NASA has at its disposal one critical tool for addressing SLS affordability concerns—competition— more specifically, competition for procurement associated with the two variants' elements. While NASA is planning to compete procurement of the advanced boosters, the agency also has opportunities to compete procurement of other upgraded elements, including the upper stage and the upper stage engine to help it achieve long-term affordability.

## RECOMMENDATIONS FOR EXECUTIVE ACTION

To provide the Congress with the necessary insight into program planning and affordability, and to decrease the risk of cost and schedule overruns, we recommend that NASA's Administrator direct the Human Exploration and Operations Mission Directorate to take the following four actions:

- NASA should develop an executable business case for SLS based on matching requirements and resources that results in a level of risk commensurate with its policies. For example, NASA could delay the planned first flight test or increase funding to allow the program to establish cost and schedule baselines for demonstration of the initial capability at the 70 percent confidence level.

- To provide decision makers with an informed basis for making investment decisions regarding the SLS program, NASA should identify a range of possible missions for each future SLS variant that includes cost and schedule estimates and plans for how those possible missions would fit within NASA's funding profile.
- To allow for a continued assessment of progress and affordability, NASA should structure each future increment of SLS capability with a total cost exceeding the $250 million threshold for designation as a major project as a separate development effort within the SLS program. In doing so, NASA should require each increment to complete both the technical and programmatic reviews required of other major development projects, per the agency's acquisition and system engineering policies.
- To promote affordability, before finalizing acquisition plans for future capability variants, NASA should assess the full range of competition opportunities and provide to the Congress the agency's assessment of the extent to which development and production of future elements of the SLS could be competitively procured.

## AGENCY COMMENTS

In written and oral comments on a draft of this report, NASA concurred with our four recommendations. NASA agreed that future increments of SLS should be structured as separate development efforts and indicated that those increments would be subject to the appropriate technical and programmatic reviews, as well as rigorous cost and schedule management. NASA also recognized the need to define missions to guide program planning and indicated it will develop plans for specific SLS missions within NASA's standard mission selection and review process.

We would expect that this process would also include cost and schedule estimates associated with each option and plans for how these possible missions would fit within NASA's funding profile. NASA acknowledged the importance of establishing a business case and cost and schedule baselines for the SLS program consistent with NASA policies and recognized that the program's schedule may have to be revised to accommodate decreased funding levels or to address technical problems. We would expect that an executable business case for the SLS program would be based on cost and schedule baselines that reflect a joint cost and schedule confidence level of 70 percent. NASA also indicated it would assess opportunities for increased competition and follow all applicable federal and NASA acquisition regulations governing competition within its contracting activities. As our recommendation indicates, NASA's assessment of competitive procurement opportunities should address development and production of all future elements of the SLS, including the upper stage, and result in a report of this current assessment to the Congress. Separately, NASA provided technical comments, which have been incorporated into the report, as appropriate.

Sincerely yours,
Cristina T. Chaplain Director
Acquisition and Sourcing Management

## APPENDIX I: SCOPE AND METHODOLOGY

In order to assess the National Aeronautics and Space Administration's (NASA) progress to conduct its first flight in 2017, we interviewed and obtained briefings and relevant documents from NASA and contractor officials. We identified and evaluated technical and programmatic issues associated with each major Space Launch System (SLS) element, including new and heritage hardware, by reviewing associated development plans and discussing relevant schedules and issues with agency officials. We also compared planned and actual progress in maturing system designs at the element and program level and evaluated how delays in element maturation and integration could affect the SLS program as a whole. We assessed NASA's risk mitigation plans for SLS to gauge the program's progress in addressing technical issues and to evaluate the potential cost and schedule impact to program milestones, including events such as delivery and flight dates. We also reviewed other technical and programmatic indicators—including progress against mass and performance goals, progress made addressing required actions from programmatic reviews at both the subsystem and vehicle level, and the status of schedule threats to the program's critical path—and assessed program status against those indicators. We assessed the status of the program's allocation of top level requirements to the different elements and the preparation of verification and validation plans for those requirements at the preliminary design review. We also evaluated the status of contracting activities for each element, including determining whether respective contracts were definitized and baselined and whether earned value information about contractor progress against planned cost and schedule was available to the program. To assess long-term affordability, we discussed long-term development plans in support of future missions with agency officials. We also reviewed the design reference architecture establishing the need for the large launch vehicle and compared the program's approach for satisfying the need to the agency's strategic plan. For purposes of assessing the cost estimate, we reviewed NASA preliminary cost estimates for the SLS, Orion, and associated ground systems programs and information related to NASA's plans for the SLS baseline cost estimate in order to determine the scope of the estimates and assessed whether that scope provided transparency into costs and enabled assessment of long-term affordability. We assessed the preliminary estimates' scope against best practices criteria outlined in GAO's cost estimating guidebook as well as NASA's own guidance and procedural requirements. We also discussed the estimate with NASA officials, including the rationale for the estimates' scope and exclusions to that scope. As NASA had not released the baseline estimate for SLS at the time of our review, we did not asses the reliability of the SLS baseline estimate. Furthermore, we evaluated the program's development and acquisition plans to compete future variants of the SLS by reviewing contract information, including any follow-on contract options, and discussing supplier availability with agency officials.

We conducted our work at locations where the SLS program is being managed and executed including NASA headquarters in Washington, D.C.; NASA's Kennedy Space Flight Center in Cape Canaveral, Fla.; NASA's Marshall Space Flight Center in Huntsville, Ala.; AeroJet Rocketdyne in Canoga Park, Calif.; Alliant Technologies in Promontory, Utah; and the Boeing Company in Huntsville, Ala.

We conducted this performance audit from June 2013 to July 2014 in accordance with generally accepted government auditing standards. Those standards require that we plan and perform the audit to obtain sufficient, appropriate evidence to provide a reasonable basis for our findings and conclusions based on our audit objectives. We believe that the evidence obtained provides a reasonable basis for our findings and conclusions based on our audit objectives.

## End Notes

[1] The White House, National Space Transportation Policy (Washington, D.C.: Nov. 21, 2013).

[2] GAO, *High-Risk Series: An Update*, GAO-13-283 (Washington, D.C.: February 2013).

[3] GAO, *GAO Cost Estimating and Assessment Guide: Best Practices for Developing and Managing Capital Program Costs,* GAO-09-3SP (Washington, D.C.: March 2009). The Guide is a compilation of best practices that federal cost estimating organizations and industry use to develop and maintain reliable cost estimates throughout the life of a government acquisition program. NASA Procedural Requirements (NPR) 7120.5E, *NASA Space Flight Program and Project Management Requirements*, § 2.4 (Aug. 14, 2012); and NASA, *2008 NASA Cost Estimating Handbook* (Washington, D.C.).

[4] Pub. L. No. 111-267, §§ 302(c)(2), 303(a)(2) (codified at 42 U.S.C. §§ 18322, 18323).

[5] NASA plans for SLS Block IA to utilize advanced boosters, Block IB an exploration upper stage, and Block II the advanced boosters and exploration upper stage. The agency has not yet determined whether it will first develop the Block IA or Block IB variant.

[6] In a two-body system, such as Earth and the sun, there are points nearby where a third object can be positioned and remain in place relative to the other two objects. These are known as Lagrange points.

[7] NASA defines the formulation phase as the identification of how the program or project supports the agency's strategic goals; the assessment of feasibility, technology, concepts, and performance of trade studies; risk assessment and possible risk mitigations and continuous risk management processes; team building, development of operations concepts and acquisition strategies; establishment of high-level requirements, requirements flow down, and success criteria; assessing the relevant industrial base/supply chain to ensure program or project success, the preparation of plans, cost estimates, budgets, and schedules essential to the success of a program or project; and the establishment of control systems to ensure performance of those plans and alignment with current agency strategies. NPR 7120.5E, § 1.3.1.a (Aug. 14, 2012). The implementation phase is defined as the execution of approved plans for the development and operation of the program or project, and the use of control systems to ensure performance to approved plans and requirements and continued alignment with the agency's strategic goals. NPR 7120.5E, §1.3.1.c (Aug. 14, 2012).

[8] NPR 7120.5E, (Aug.14, 2012).

[9] NPR 7120.5E, Appendix A (Aug. 14, 2012).

[10] NPR 7120.5E, § 2.4 (Aug. 14, 2012).

[11] NPR 7120.5E, Table 2-5 (Aug. 14, 2012).

[12] NPR 7120.5E, § 2.4 (Aug. 14, 2012).

[13] GAO, *Best Practices: Capturing Design and Manufacturing Knowledge Early Improves Acquisition Outcomes*, GAO-03-645T (Washington, D.C.: Jul. 15, 2002).

[14] GAO, *NASA: Implementing a Knowledge-Based Acquisition Framework Could Lead to Better Investment Decisions and Project Outcomes*, GAO-06-218 (Washington, D.C.: Dec. 21, 2005), and *NASA: Agency Has Taken Steps Toward Making Sound Investment Decisions for Ares I but Still Faces Challenging Knowledge Gaps*, GAO-08-51 (Washington, D.C.: Oct. 31, 2007).

[15] NPR 7120.5E, Table 2-5 (Aug. 14, 2012).

[16] GAO, GAO-03-645T; GAO-06-218; *Best Practices: Using a Knowledge-based Approach to Improve Weapon Acquisition*, GAO-04-386SP (Washington, D.C.: Jan. 1, 2004); and NASA, *Systems Engineering Handbook,* NASA/SP-2007-6105 Rev1 (Washington, D.C.: December 2007).

[17] Mass is a measurement of how much matter is in an object. It is related to an object's weight, which is mathematically equal to mass multiplied by acceleration due to gravity. Margin is the spare amount of mass or power allowed or given for contingencies or special situations.

[18] GAO, *NASA: Assessments of Selected Large-Scale Projects*, GAO-14-338SP (Washington, D.C.: Apr. 15, 2014).

[19] NASA's procedural requirements require Mission Directorates to plan and budget programs and projects with an estimated life-cycle cost greater than $250 million based on a 70 percent Joint Cost and Schedule Confidence Level (JCL), or at a different level as approved by the Decision Authority, which for SLS is the NASA Associate Administrator. Any JCL approved by the Decision Authority at less than 70 percent must be justified and documented. The JCL is a quantitative probability analysis that requires the project to combine its cost, schedule, and risks into a complete quantitative picture to help assess whether the project will be successfully completed within cost and on schedule. NPR 7120.5E, §§2.4.4, 2.4.4.1, 2.4.3.2 (Aug. 14, 2012).

[20] GAO, GAO-04-386SP and *Best Practices: Better Matching of Needs and Resources Will lead to Better Weapon System Outcomes*, GAO-01-288 (Washington, D.C.: Mar. 8, 2001).

[21] GAO-04-386SP and GAO-01-288.

[22] GAO, *Space Transportation: Status of the X-33 Reusable Launch Vehicle Program*, GAO/NSIAD-99-176 (Washington, D.C.: Aug. 11, 1999); *National Aero-Space Plane: Restructuring Future Research and Development Efforts*, GAO/NSIAD-93-71, (Washington, D.C.: Dec. 3, 1992); and *NASA: Constellation Program Cost and Schedule Will Remain Uncertain Until a Sound Business Case Is Established*, GAO-09-844 (Washington, D.C.: Aug. 26, 2009).

[23] GAO, *Defense Acquisitions: Key Decisions to Be Made on Future Combat System*, GAO-07-376 (Washington, D.C.: Mar. 15, 2007); *Defense Acquisitions: Improved Business Case Key for Future Combat System's Success*, GAO-06-564T (Washington, D.C.: Apr. 4, 2006); GAO-06-218; and *NASA's Space Vision: Business Case for Prometheus 1 Needed to Ensure Requirements Match Available Resources*, GAO-05-242 (Washington, D.C.: Feb. 28, 2005).

[24] Engineering drawings are considered to be a good measure of the demonstrated stability of a product's design because the drawings represent the language used by engineers to communicate to the manufacturers the details of a new product design— what it looks like, how its components interface, how it functions, how to build it, and what critical materials and processes are required to fabricate and test it. Once the design of a product is finalized, the drawing is "releasable."

[25] NASA, Goddard Procedural Requirements 7120.7 (May 4, 2008).

[26] GAO, *NASA: Assessments of Selected Large-Scale Projects*, GAO-13-276SP (Washington, D.C.: Apr. 17, 2013).

[27] GAO, *NASA: Assessments of Selected Large-Scale Projects*, GAO-12-207SP, (Washington, D.C.: Mar. 1, 2012), and NASA Office of Inspector General, *NASA's Challenges to Meeting Cost, Schedule, and Performance Goals*, IG-12-021, (Washington, D.C.: Sept. 27, 2012).

[28] NASA Federal Acquisition Regulation Supplement (NFS) § 1843.7005(a).

[29] NFS § 1843.7002(a).

[30] GAO-09-3SP and NPR 7120.5E, § 2.2.8 (Aug. 14, 2012).

[31] GAO, *Defense Contract Management: DOD's Lack of Adherence to Key Contracting Principles on Iraq Oil Contract Put Government Interests at Risk*, GAO-07-839 (Washington, D.C.: Jul. 31, 2007); *Defense Contracting: Use of Undefinitized Contract Actions Understated and Definitization Time Frames Often Not Met*, GAO-07-559 (Washington, D.C.: Jun. 19, 2007); and *Rebuilding Iraq: Fiscal Year 2003 Contract Award Procedures and Management Challenges*, GAO-04-605 (Washington, D.C.: Jun. 1, 2004).

[32] See, for example, GAO, *NASA: Earned Value Management Implementation across Major Spaceflight Projects Is Uneven*, GAO-13-22, (Washington, D.C.: Nov. 19, 2012); *Defense Acquisitions: Missile Defense Program Instability Affects Reliability of Earned Value Management Data*, GAO-10-676 (Washington, D.C.: Jul. 14, 2010); and GAO-09-3SP.

[33] Office of Management and Budget, *OMB Circular A-11, Supplement for Part 7: Capital Programming Guide* (Washington, D.C.: July 2013).

[34] NPR 7120.5 E (Aug. 14, 2012).

[35] GAO-09-3SP; GAO-04-386SP; and GAO-01-288.

[36] GAO, *Tactical Aircraft: F-22A Modernization Program Faces Cost, Technical, and Sustainment Risks*, GAO-12-447 (Washington D.C.: May 2, 2012).

[37] GAO, *NASA: Actions Needed to Improve Transparency and Assess Long-Term Affordability of Human Exploration Programs*, GAO-14-385, (Washington D.C.: May 8, 2014).

[38] GAO-09-3SP; NPR 7120.5E § 2.4(Aug. 14, 2012); *2008 NASA Cost Estimating Handbook; and* GAO-14-385.

[39] National Space Transportation Policy. (Nov. 21, 2013).

[40] See, for example, GAO, *Federal Contracting: Noncompetitive Contracts Based on Urgency Need Additional Oversight*, GAO-14-304 (Washington, D.C.: Mar. 26, 2014); *Defense Contracting: Actions Needed to Increase Competition*, GAO-13-325 (Washington, D.C.: Mar. 28, 2013); and *Federal Contracting: Opportunities Exist to Increase Competition and Assess Reasons When Only One Offer Is Received*, GAO-10-833 (Washington, D.C.: Jul. 26, 2010).

[41] Pub. L. No. 98-369, § 2711, 98 Stat. 494, 1175; and Federal Acquisition Regulation § 6.101(a).

In: Human Space Exploration
Editor: Darrel Gibbs

ISBN: 978-1-63482-172-8


*Chapter 3*

# NASA: Human Space Exploration Programs Face Challenges. Statement of Cristina T. Chaplain, Director, Acquisition and Sourcing Management, Government Accountability Office. Hearing on "An Update on the Space Launch System and Orion: Monitoring the Development of the Nation's Deep Space Exploration Capabilities"*

## Why GAO Did This Study

NASA is undertaking a trio of closely related programs to continue human space exploration beyond low-Earth orbit: the SLS vehicle; the Orion capsule, which will launch atop the SLS and carry astronauts; and GSDO, the supporting ground systems. As a whole, the efforts represent NASA's largest exploration investment over the next decade, approaching $23 billion, to demonstrate initial capabilities.

---

* This is an edited, reformatted and augmented version of testimony presented December 10, 2014 before the House Committee on Science, Space and Technology, Subcommittee on Space.

In May 2014, GAO found that NASA's preliminary life-cycle cost estimates for human exploration were incomplete and recommended that NASA establish life-cycle cost and schedule baselines for each upgraded block of SLS, Orion, and GSDO; NASA partially concurred. In July 2014, GAO issued a report on SLS's progress toward its first test flight and recommended that NASA match SLS's resources to its requirements and define specific missions beyond the second test flight, among other actions. NASA concurred with these recommendations.

This testimony is based on GAO's May 2014 report (GAO-14-385), July 2014 report (GAO-14-631), and ongoing audit work related to SLS and Orion. It discusses NASA's efforts to match resources to requirements for the SLS program and developmental challenges facing the SLS and Orion programs. To conduct this work, GAO reviewed relevant design, development, cost, and schedule documents and interviewed program officials.

## What GAO Found

In 2014, GAO reported on a number of issues related to the National Aeronautics and Space Administration's (NASA) human exploration programs: the Space Launch System (SLS) vehicle, the Orion Multi-Purpose Crew Vehicle (Orion), and the Ground Systems Development and Operations (GSDO). For example, in July 2014, GAO found that NASA had not matched resources to requirements for the SLS program and was pursuing an aggressive development schedule—a situation compounded by the agency's reluctance to request funding commensurate with the program's needs. In August 2014, NASA established formal cost and schedule baselines for the SLS program at the agency-required 70 percent joint cost and schedule confidence level (JCL), which satisfied one recommendation from GAO's July 2014 report. The JCL is a calculation NASA uses to estimate the probable success of a program meeting its cost and schedule targets. To satisfy the 70 percent JCL requirement, the SLS program delayed its committed launch readiness date for its first test flight from December 2017 to November 2018. The program is still pursuing December 2017 as an internal goal, or target date, for the test flight, even though NASA calculated the JCL associated with launching SLS on this date at 30 percent. Moreover, neither the Orion nor GSDO program expects to be ready for the December 2017 launch date. With these programs likely unable to meet the December 2017 date, NASA risks exhausting limited human exploration resources to achieve an accelerated SLS program schedule when those resources may be needed to resolve challenges on other human exploration programs.

**NASA's Target and Baseline Launch Readiness Dates and Associated Confidence Levels for Human Space Exploration Programs**

| | Target date | Confidence level for target date | Committed date | Confidence level for committed date |
|---|---|---|---|---|
| Space Launch System | December 2017 | 30% | November 2018 | 70% |
| Ground Systems Development and Operations | June 2018 | 30% | November 2018 | 80% |
| Orion Multi-Purpose Crew Vehicle[a] | TBD | TBD | TBD | TBD |

Source: GAO analysis of NASA data. | GAO-15-248T

[a]Orion has yet to establish formal cost and schedule baseline commitments.

In addition, GAO's ongoing work has found that the Orion program is facing significant technical and funding issues. Orion just completed its first test flight, and data from this flight is required to address several risks that must be resolved before the second test flight in 2021 because they represent risks to crew safety. For example, during parachute testing, NASA discovered that when only two of the three main parachutes are deployed, they begin to swing past each other creating a "pendulum" effect. This effect could cause the capsule to increase speed and to hit the water at an angle that may damage the capsule, thereby endangering the crew. In addition, data from the test is necessary to inform NASA's design solution to address heat shield cracking issues, which NASA has been working to resolve since August 2013. The heat shield is integral to crew safety during re-entry.

Chairman Palazzo, Ranking Member Edwards, and Members of the Subcommittee:

I am pleased to be here today to discuss our work on the Space Launch System (SLS) and National Aeronautics and Space Administration's (NASA) other human exploration programs. SLS is NASA's first exploration-class launch vehicle in over 40 years. It is being developed to launch astronauts and carry cargo into space, beyond low-Earth orbit. SLS progress cannot be fully discussed without considering the progress of the Orion Multi-Purpose Crew

Vehicle (Orion) and Ground Systems Development and Operations (GSDO) programs. The Orion program is developing a capsule that will launch atop the SLS and carry the astronauts, and the GSDO program is developing systems and infrastructure to support such activities as assembly, test, and launch of the SLS and Orion. As a whole, these efforts may cost nearly $23 billion to demonstrate initial capabilities, including the first planned SLS flight in 2018, the ground systems for that effort, and the first two Orion flights currently planned for fiscal years 2018 and 2021. This amount represents a significant portion of NASA's planned budget for major projects during that period and also a significant portion of government-wide launch-related research and development funding. As we have reported, any cost or schedule overrun on NASA's largest, most complex projects— including SLS, Orion, and GSDO—could have a ripple effect on the portfolio and has the potential to postpone, or even cancel altogether, projects in earlier development stages.[1] Given the expensive nature of developing space systems for human exploration, in today's constrained government budget environment, it is essential that NASA manage the acquisition of these systems as efficiently and effectively as possible.

GAO has designated NASA's management of acquisitions as a high-risk area for more than two decades in view of persistent cost growth and schedule slippage in the majority of its major projects. NASA's attempts to develop systems capable of transporting humans to space since the development of the Space Shuttle have been unsuccessful. For example, prior development programs, the most recent being the Constellation program, were canceled in the face of acquisition problems and funding-related issues. While the agency has made progress in recent years in reducing risk on smaller-scale, less complex projects, demonstrating that this progress can be translated to larger, more complex projects, such as SLS and Orion, is more challenging yet exceedingly important.

We recently issued two reports on NASA's current human exploration program.[2] In addition, we regularly review SLS and Orion as part of our annual review of major NASA programs and projects.[3] My remarks today are based on these reports and our ongoing work. Specifically, I will discuss NASA's efforts to establish baselines for the SLS program based on matching cost and schedule resources to requirements,

developmental challenges facing the SLS and Orion programs, and shortcomings in NASA's planning and cost estimates for the SLS, Orion, and GSDO programs. To conduct this work, we reviewed relevant design, development, cost, and schedule documents; interviewed program officials;

and evaluated SLS and Orion program actions using acquisition and cost estimating best practices. More information on our scope and methodology is available in our related GAO products.

The work that supports this statement was performed in accordance with generally accepted government auditing standards. Those standards require that we plan and perform the audit to obtain sufficient, appropriate evidence to provide a reasonable basis for our findings and conclusions based on our audit objectives. We believe that the evidence obtained provides a reasonable basis for our findings and conclusions based on our audit objectives.

## Summary

In 2014, we found that NASA had not matched cost and schedule resources to requirements for the SLS program and was pursuing an aggressive development schedule. This situation, in turn, was compounded by the agency's reluctance to request funding in line with the program's needs. In addition, we found that the agency's preliminary life-cycle cost estimates for human exploration were incomplete. Based on these findings, in July 2014, we recommended, among other things, that NASA develop baselines for SLS that match cost and schedule resources to requirements and result in a level of risk in line with its policies as well as develop improved life-cycle cost estimates. NASA concurred with our findings and recommendations. In August 2014, NASA completed the review of the SLS program that sets formal cost and schedule baselines and, in doing so, delayed the first test flight to relieve schedule pressure and allow additional time to address design challenges. However, some of the concerns we raised about the cost estimates, mission requirements, and long-term affordability remain. In addition, our ongoing work has found that the three human exploration programs are pursuing inconsistent and unrealistic schedule goals and that the Orion program is facing significant technical and funding issues that may affect NASA's overall schedule for its human exploration programs.

## Background

The National Aeronautics and Space Administration Authorization Act of 2010 directed NASA to, among another things, develop a Space Launch

System as a follow-on to the Space Shuttle and as a key component in expanding human presence beyond low-Earth orbit. To that end, NASA plans to incrementally develop three progressively more capable SLS launch vehicles—70-, 105-, and 130-metric ton (mt) variants. When complete, the 130-mt vehicle is expected to have more launch capability than the Saturn V vehicle, which was used for Apollo missions, and be significantly more capable than any recent or current launch vehicle. The act also directed NASA to prioritize the core elements of SLS with the goal of operational capability not later than December 2016.[4] NASA negotiated an extension of that date, to December 2017, based on the agency's initial assessment of the tasks associated with developing the new launch vehicle, and has subsequently committed to a launch readiness date of November 2018.

In 2011, NASA formally established the SLS program. To fulfill the direction of the 2010 act, the agency plans to develop the three SLS launch vehicle capabilities, complemented by Orion, to transport humans and cargo into space. The first version of the SLS that NASA is developing is a 70-mt launch vehicle known as Block I. NASA has committed to conduct two test flights of the Block I vehicle—the first in 2018 and the second in 2021. The vehicle is scheduled to fly an uncrewed Orion some 70,000 kilometers beyond the moon during the first test flight, known as Exploration Mission-1 (EM-1), and to fly a second mission known as Exploration Mission-2 (EM-2) beyond the moon to further test performance with a crewed Orion vehicle. After 2021, NASA intends to build 105- and 130-mt launch vehicles, known respectively as Block IA/B and Block II, which it expects to use as the backbone of manned spaceflight for decades.[5] NASA anticipates using the Block IA/B vehicles for destinations such as near-Earth asteroids and LaGrange points and the Block II vehicles for eventual Mars missions.[6]

Space launch vehicle development efforts are high risk from technical, programmatic, and oversight perspectives. The technical risk is inherent for a variety of reasons including the environment in which they must operate, complexity of technologies and designs, and limited room for error in the fabrication and integration process. Managing the development process is complex for reasons that go well beyond technology and design. For instance, at the strategic level, because launch vehicle programs can span many years and be very costly, programs often face difficulties securing and sustaining funding commitments and support.

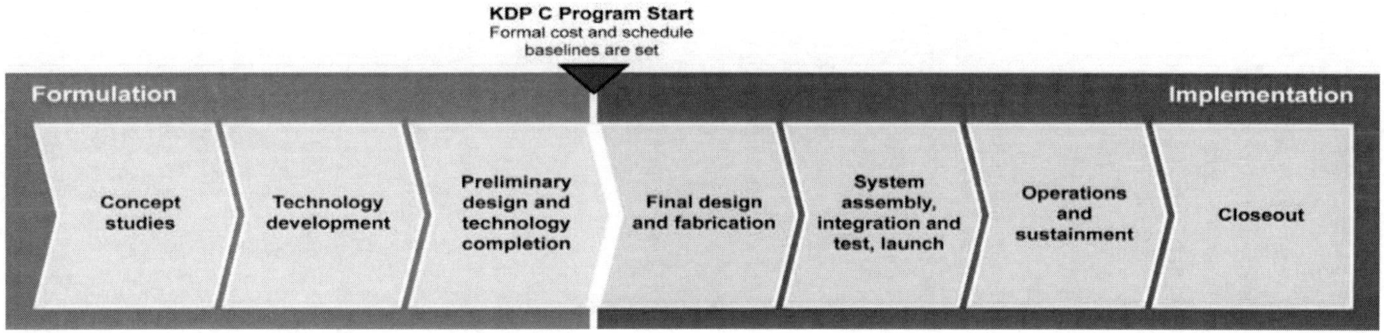


Source: NASA data and GAO analysis. | GAO-15-248T.

Figure 1. Key Decision Point (KDP)-C in NASA Development Life Cycle.

At the program level, if the lines of communication between engineers, managers, and senior leaders are not clear, risks that pose significant threats could go unrecognized and unmitigated. If there are pressures to deliver a capability within a short period of time, programs may be incentivized to overlap development and production activities or delete tests, which could result in late discovery of significant technical problems that require more money and ultimately much more time to address. For these reasons, it is imperative that launch vehicle development efforts adopt disciplined practices and lessons learned from past programs.

Best practices for acquisition programs indicate that establishing baselines that match cost and schedule resources to requirements and rationally balancing cost, schedule, and performance is a key step in establishing a successful acquisition program.[7] Our work has also shown that validating this match before committing resources to development helps to mitigate the risks inherent in NASA's programs.[8] We have reported that within NASA's acquisition life cycle, resources should be matched to requirements at key decision point (KDP)-C, the review that commits the program to formal cost and schedule baselines and marks the transition from the formulation phase into the implementation phase, as seen in figure 1 below.[9] The SLS program completed its KDP-C review in August 2014, GSDO completed its KDP-C review in September 2014, and the KDP-C review for Orion is currently scheduled for May 2015.

## NASA Delayed SLS Launch Date to Better Match Resources to Requirements

NASA has taken positive steps to address specific concerns we raised in July 2014 regarding aggressive schedules and insufficient funding by establishing the SLS program's committed launch readiness date as November 2018—almost a year later than originally planned. Specifically, we reported in July 2014 that NASA had yet to establish baselines that matched the SLS program's cost and schedule resources with the requirement to develop the SLS and launch the first flight test in December 2017 at the required confidence level of 70 percent.[10] NASA policy generally requires a 70 percent joint confidence level—a calculation NASA uses to estimate the probable success of a program meeting its cost and schedule targets—for a program to proceed with final design and fabrication. At the time of our July 2014 report,

NASA had delayed its review to formally commit the agency to cost and schedule baselines for SLS from October 2013, as the agency considered future funding plans for the program. At that time, the agency's funding plan for SLS was insufficient to match requirements to resources for the December 2017 flight test at the 70 percent joint confidence level and the agency's options for matching resources to requirements were largely limited to increasing program funding, delaying the schedule, or accepting a reduced confidence level for the initial flight test. We have previously reported that it is important for NASA to budget projects to appropriate confidence levels, as past studies have linked cost growth to insufficient reserves, poorly phased funding profiles, and more generally, optimistic estimating practices.

We found that NASA's proposed funding levels had affected the SLS program's ability to match requirements to resources since its inception. NASA has requested relatively consistent amounts of funding of about $1.4 billion each year since 2012. According to agency officials, the program has taken steps to operate within that flat funding profile, including streamlining program office operations and asking each contractor to identify efficiencies in its production processes. Even so, according to the program's own analysis, going into the agency review to formally set baselines, SLS's top risk was that the current planned budget through 2017 would be insufficient to allow the SLS as designed to meet the EM-1 flight date. The SLS program office calculated the risk associated with insufficient funding through 2017 as 90 percent likely to occur; furthermore, it indicated the insufficient budget could push the December 2017 launch date out 6 months and add some $400 million to the overall cost of SLS development. The cost risk was considerably greater than $400 million in the past, but according to program officials they were able to reduce the affect due to receiving more funding than requested in fiscal years 2013 and 2014. Similarly, our ongoing work on human exploration programs has found that the Orion program is currently tracking a funding risk that the program could require an additional $560 to $840 million to meet the December 2017 EM-1 flight date. However, the agency has yet to complete the review that sets formal cost or schedule baselines for the Orion program. At this time, we have not conducted enough in-depth work on the GSDO program to comment on any specific risks the program is tracking.

In our July 2014 report we recommended, among other things, that NASA develop baselines for SLS based on matching cost and schedule resources to requirements that would result in a level of risk commensurate with its policies. NASA concurred with our findings and recommendations. In August 2014, NASA established formal cost and schedule baselines for the SLS program at the 70 percent joint confidence level for a committed launch readiness date of November 2018. Nevertheless, the program plans to continue to pursue an initial capability of SLS by December 2017 as an internal goal and has calculated a joint cost and schedule confidence level of 30 percent associated with that date.

As illustrated by table 1 below, the SLS and GSDO programs are pursuing ambitious and varying target dates for the EM-1 test flight. In addition, the Orion program is currently tracking and reporting to December 2017. The agency acknowledges differences in the target dates the programs are pursuing and has indicated that it will develop an integrated target launch date after all three systems hold their individual critical design reviews.

**Table 1. NASA's Target and Baseline Launch Readiness Dates and Associated Confidence Levels for Human Spaceflight Programs**

| | Target date | Confidence level for target date | Committed date | Confidence level for committed date |
|---|---|---|---|---|
| Space Launch System | December 2017 | 30% | November 2018 | 70% |
| Ground Systems Development and Operations | June 2018 | 30% | November 2018 | 80% |
| Orion Multi-Purpose Crew Vehicle[a] | TBD | TBD | TBD | TBD |

Source: GAO analysis of NASA data. | GAO-15-248T.

[a]Orion has not yet established formal cost and schedule baseline commitments.

The SLS program has assigned a low confidence level—30 percent—associated with meeting the program's internal target date of December 2017. Even if SLS does meet that goal, however, it is unlikely that both Orion and GSDO will achieve launch readiness by that point. For example, the GSDO program only has a 30 percent confidence level associated with a later June 2018 date. Additionally, the Orion program is currently behind its planned schedule and is facing significant technical risks and officials indicated that the program will not achieve launch readiness by December 2017. The Orion program has submitted a schedule to NASA headquarters that indicates the program is now developing plans for a September 2018 EM-1 launch, though that date is preliminary until the program establishes official cost and schedule baselines now planned for May 2015. With the Orion and GSDO programs likely unable to meet the December 2017 date, NASA risks exhausting limited human exploration resources to achieve an aggressive SLS program schedule when those resources may be needed to resolve other issues within the human exploration effort. In other work, we have reported that in pursuing internal schedule goals, some programs have exhausted cost reserves, which has resulted in the need for additional funding to support the agency baseline commitment date once the target date is not achieved.[11]

NASA's urgency to complete development and demonstrate a human launch capability as soon as possible is understandable. The United States has lacked the ability to launch humans into space since the last flight of the Space Shuttle in July 2011 and the initial goal from Congress was that NASA demonstrate a new human launch capability by 2016. Also, the SLS and GSDO programs have already slipped their committed launch readiness dates to November 2018, and Orion appears likely to follow suit. While these delays were appropriate actions on the agency's part to reduce risk, their compounding effect could have impacts on the first crewed flight—EM-2—currently scheduled for 2021.

## NASA's SLS and Orion Programs Are Making Progress, but the Orion Program Is Facing Technical Challenges

We reported in July 2014 that NASA's metrics indicated the SLS program was on track to meet many of its design goals for demonstrating the initial capability of SLS. However, we found that the development of the core stage—SLS's fuel tank and structural backbone—represents the critical path of activities that must be completed to maintain the program's schedule as a whole.

The core stage development had an aggressive schedule in order to meet the planned December 2017 first test flight. For example, the core stage had threats of nearly 5 months to its schedule due to difficulty acquiring liquid oxygen fuel lines capable of meeting SLS operational requirements. The aggressiveness of, and therefore the risk associated with the core stage schedule was reduced when the agency delayed its commitment for initial capability of SLS until November 2018. With SLS continuing to pursue a target date of December 2017, however, the aggressive core stage schedule remains a risk.

Further, we reported that the program faced challenges integrating heritage hardware, which was designed for less stressful operational environments, into the SLS design. We found that these issues were not significant schedule drivers for the program as each had, and continues to have, significant amounts of schedule reserve to both the target and agency baseline commitment dates for launch readiness.

The Orion program just completed its first experimental test flight—EFT-1. This flight tested Orion systems critical to crew safety, such as heat shield performance, separation events, avionics and software performance, attitude control and guidance, parachute deployment, and recovery operations. According to NASA, the data gathered during the flight will influence design decisions and validate existing computer models. Data from this flight are required to address several significant risks that the Orion program is currently tracking that must be addressed before humans can be flown on Orion. Specifically, our ongoing work indicates that the Orion program passed its preliminary design review—a review that evaluates the adequacy of cost schedule and technical baselines and whether the program is ready to move forward—in August 2014 by meeting the minimum standards for all 10 success criteria. For 7 of the 10 success criteria, however, review officials highlighted known issues that could compromise Orion's success. Specifically, the review officials noted concerns about several unresolved design risks, including technical challenges with the parachute system and heat shield. For example, during parachute testing, NASA discovered that when only two of the three main parachutes are deployed, they begin to swing past each other creating a "pendulum" effect. This effect could cause the capsule to increase speed and to hit the water at an angle that may damage the capsule thereby endangering the crew.

Further, NASA faces choices between differing design solutions to resolve cracking issues discovered during manufacturing of the heat shield that protects the capsule during re-entry. Program officials plan to make a decision prior to the program's critical design review, based on additional testing and analysis, about how to resolve these risks with a goal of limiting design changes to the capsule's structure. Both the parachute and heat shield challenges must be resolved before EM-2 because each represents a significant risk to crew safety. Significant cost and schedule impacts could result if a redesign is required to address any of these unresolved design risks.

## NASA'S HUMAN EXPLORATION PROGRAMS' LONG-TERM MISSIONS AND AFFORDABILITY ARE UNCERTAIN

NASA has yet to address our concerns regarding mission planning or life-cycle cost estimates. NASA has not yet defined specific mission requirements for any variant of the SLS. The two currently scheduled flights are developmental test flights designed to demonstrate and test the capabilities of the 70-mt launch vehicle and the capability of the core stage in particular. Office of Management and Budget guidance indicates that agencies should develop long-range objectives, supported by detailed budgets and plans that identify the agency's performance gaps and the resources needed to close them.[12] With mission requirements unspecified, NASA has not yet finalized plans for the next step in evolving the SLS and risks investing limited available resources in systems and designs that are not yet needed and missing opportunities to make early investments in developing systems that may be needed in the future.

According to agency officials, beyond the two scheduled test flights, future mission destinations remain uncertain. In the absence of specific mission requirements, officials indicated the SLS program is developing current and future variants based on top-level requirements derived from NASA's Design Reference Architectures for conducting missions in line with the agency's strategic plan. NASA's 2014 strategic plan, for example, identifies sending humans to Mars as one of the agency's long-term goals; in turn, the agency's Mars Design Reference Architecture indicates that multiple missions using a vehicle with a lift capability of about 130-mt will be necessary to support that goal. We recommended based on these findings that NASA define a range of possible missions beyond the second test flight and introduce increased competition in the acquisition of hardware needed for future variants to reduce long-term costs.[13] The agency concurred with our recommendations, but has not yet taken specific actions to address our concerns.

**Table 2. Costs Included in the Scope of Baseline or Preliminary Cost Estimate**

| | Agency baseline commitment (costs) | System development (including establishing manufacturing and test facilities) | Exploration Mission 1 (including 3 months of post-flight data analysis) | Exploration Mission 2 (including 3 months of post-flight data analysis) | Development of future vehicles | Future manufacturing, operations, and support costs |
|---|---|---|---|---|---|---|
| Space Launch System (SLS) | $9.7 billion | Yes | Yes | No | No | No |
| Ground Systems Development and Operations | $2.8 billion | Yes | Yes | No | No | No |
| Orion Multi-Purpose Crew Vehicle | TBD Preliminary range estimate of $8.5 billion to $10.3 billion | Yes | Yes | Yes | No | No |

Source: GAO analysis of NASA data. | GAO-15-248T

The long-term affordability of the human exploration programs are also uncertain, as we found in May 2014, because NASA's cost estimates for the programs do not provide any information about the longer-term, life-cycle costs of developing, manufacturing, and operating the launch vehicles.[14] For example, as illustrated in table 2 below, NASA's baseline estimate for SLS does not cover program costs after EM-1 or costs to design, develop, build, and produce the 105- or 130-mt variants. Though the subsequent variants will evolve from the first variant, they each represent substantial, challenging development efforts and will require billions of more dollars to complete. For example, the 105-mt vehicle will require development of a new upper stage and upper stage engine or the development of advanced boosters, either of which will be significant efforts for the program.

Based on the tenets of widely accepted best practices for cost estimation, as well as NASA's own requirements and guidance regarding life-cycle costs, in May 2014 we recommended that NASA establish a separate cost and schedule baseline for the SLS program for work required to support EM-2.[15] Additionally, we recommended that NASA establish life-cycle cost and schedule baselines, or at least provide minimum and maximum ranges, for each upgraded block of SLS, Orion, and associated ground support. NASA partially concurred with our recommendations, stating that their current approach for establishing separate baselines and estimates for the SLS, Orion, and GSDO programs met the intent of our recommendations and agreed to report cost estimates for the future SLS capabilities annually via the agency budget submission until key requirements are defined and baselines can be established. We disagreed and stated that establishing cost and schedule baselines at the program level was unlikely to provide the detail necessary to monitor the progress of future blocks of SLS, each of which will in essence constitute a separate development project within the SLS program, and that budget requests neither offer all the same information as life-cycle cost estimates nor are necessarily linked to an established baseline that indicates how much NASA expects to invest to develop, operate, and sustain a capability over the long term.

In conclusion, by delaying the committed launch readiness date and establishing funding levels at a 70 percent confidence level, NASA has improved the SLS program's overall risk posture. We are concerned, however, that the program continues to pursue the overly ambitious goal of a December 2017 launch date. It is important to note at this point that the SLS, Orion, and GSDO programs are intrinsically linked. None of the three can satisfy NASA's human exploration goals on its own, and cost overruns or delays in any single program, such as the significant funding and technical issues now facing the Orion program, will directly affect the others. Without a realistic integrated flight date guiding the efforts of all three programs, and meaningful reporting of progress, insight into the progress of NASA's human exploration portfolio and the agency's ability to make informed management decisions regarding the allocation of resources across the three programs is limited. Further, NASA's plans for human exploration beyond SLS's second flight in 2021 remain unclear. Until long-term missions are finalized, the agency will lack clear definition in its plans to move forward. This will in turn affect the agency's acquisition planning and any efforts to incorporate increased competition. Furthermore, without complete life-cycle cost estimates for all three programs, and their planned variants, the agency's ability to make important decisions about the affordability of the program in the context of the agency's overall budget and competing priorities is limited.

Chairman Palazzo, Ranking Member Edwards, and Members of the Subcommittee, this completes my prepared statement. I would be pleased to respond to any questions that you may have at this time.

## End Notes

[1] GAO, *NASA: Assessment of Large Scale Projects*, GAO-14-338SP (Washington, D.C.: Apr. 15, 2014).

[2] GAO, *Space Launch System: Resources Need to be Matched to Requirements to Decrease Risk and Support Long Term Affordability,* GAO-14-631 (Washington, D.C.; July 23, 2014); and *NASA: Actions Needed to Improve Transparency and Assess Long-Term Affordability of Human Exploration Programs*, GAO-14-385 (Washington, D.C.: May 8, 2014).

[3] The explanatory statement of the House Committee on Appropriations accompanying the Omnibus Appropriations Act, 2009 required GAO to prepare project status reports on selected large-scale NASA programs, projects, or activities. 155 Cong. Rec. H1653, 1824-25 (daily ed., Feb. 23, 2009).

[4] Pub. L. No. 111-267, §§ 302(c)(2), 303(a)(2) (codified at 42 U.S.C. §§ 18322, 18323).

[5] NASA plans for SLS Block IA to utilize advanced boosters, Block IB an exploration upper stage, and Block II the advanced boosters and exploration upper stage. The agency has not yet determined whether it will first develop the Block IA or Block IB variant.

[6] In a two-body system, such as Earth and the sun, there are points nearby where a third object can be positioned and remain in place relative to the other two objects. These are known as Lagrange points.

[7] GAO-04-386SP and GAO-01-288.

[8] GAO, *Defense Acquisitions: Key Decisions to Be Made on Future Combat System*, GAO-07-376 (Washington, D.C.: Mar. 15, 2007); *Defense Acquisitions: Improved Business Case Key for Future Combat System's Success*, GAO-06-564T (Washington, D.C.: Apr. 4, 2006); *NASA: Implementing a Knowledge-Based Acquisition Framework Could Lead to Better*

*Investment Decisions and Project Outcomes*, GAO-06-218 (Washington, D.C.: Dec. 21, 2005); and *NASA's Space Vision: Business Case for Prometheus 1 Needed to Ensure Requirements Match Available Resources*, GAO-05-242 (Washington, D.C.: Feb. 28, 2005).

[9] GAO, GAO-06-218 and *NASA: Agency Has Taken Steps Toward Making Sound Investment Decisions for Ares I but Still Faces Challenging Knowledge Gaps*, GAO-08-51 (Washington, D.C.: Oct. 31, 2007).

[10] NASA's procedural requirements require Mission Directorates to plan and budget programs and projects with an estimated life-cycle cost greater than $250 million based on a 70 percent Joint Cost and Schedule Confidence Level (JCL), or at a different level as approved by the Decision Authority, which for SLS is the NASA Associate Administrator. Any JCL approved by the Decision Authority at less than 70 percent must be justified and documented. The JCL is a quantitative probability analysis that requires the project to combine its cost, schedule, and risks into a complete quantitative picture to help assess whether the project will be successfully completed within cost and on schedule. NPR 7120.5E, §§2.4.4, 2.4.4.1, 2.4.3.2 (Aug. 14, 2012).

[11] GAO-14-338SP

[12] Office of Management and Budget, *OMB Circular A-11, Supplement for Part 7: Capital Programming Guide* (Washington, D.C.: July 2013).

[13] GAO-14-631

[14] The Orion program has not yet established formal cost and schedule baseline commitments.

[15] GAO-14-385.

In: Human Space Exploration
Editor: Darrel Gibbs

ISBN: 978-1-63482-172-8


*Chapter 4*

# STATEMENT OF WILLIAM H. GERSTENMAIER, ASSOCIATE ADMINISTRATOR FOR HUMAN EXPLORATION AND OPERATIONS, NASA. HEARING ON "AN UPDATE ON THE SPACE LAUNCH SYSTEM AND ORION: MONITORING THE DEVELOPMENT OF THE NATION'S DEEP SPACE EXPLORATION CAPABILITIES"*

Mr. Chairman and Members of the Subcommittee, thank you for this opportunity to testify before you about the progress we are making in developing NASA's next generation of human deep-space exploration vehicles: the Orion crew vehicle and the heavy-lift Space Launch System (SLS).

Orion and SLS are part of a larger exploration architecture that will enable multiple missions and destinations over the next few decades, including human missions to asteroids and Mars. Affordable manufacturing and operating costs will be important to the sustainability of SLS and Orion. NASA's near-term strategy for exploration has several interlocking

* This is an edited, reformatted and augmented version of a statement presented December 10, 2014 before the House Committee on Science, Space and Technology, Subcommittee on Space.

components: using the unique environment of the International Space Station (ISS) to conduct the research and technology demonstrations necessary to keep our crews safe and productive on long-duration spaceflights; partnering with commercial entities to develop the capacity to transport cargo and crew affordably to low-Earth orbit (LEO); working in collaboration with NASA's Science, Space Technology, and Aeronautics Research Mission Directorates to better understand exploration destinations, improve our ability to work there, and understand aerodynamics at Mars and upon Earth return; continuing to work with international partners to jointly explore our solar system; and moving outward to deep space with Orion and the SLS to take us there.

Orion and the SLS are foundational capabilities for the implementation of our integrated human and robotic exploration strategy. We will travel beyond LEO to the proving ground of cis-lunar space where we will expand and test our capabilities with a series of crewed missions with SLS and Orion, including a mission to rendezvous with a redirected asteroid in lunar orbit. The Orion and SLS systems are designed to allow us to build the skills and capabilities necessary for deep-space human exploration in the proving ground of space around the Moon. These steps will build the foundation for further deep-space exploration. With the technologies and techniques we develop, we will enable expeditions to multiple destinations, allowing us to access many other destinations and ultimately pioneer Mars. We are laying the groundwork for extending human presence in the solar system. Conceived in coordination with our international partners, this strategy maintains America's role as the world's leader and foundational partner in space exploration.

## EXPLORATION FLIGHT TEST-1

Last week's successful Exploration Flight Test-1 (EFT-1) is a critical milestone on our journey to Mars and serves as the pathfinder to validate approaches to space systems development. The test demonstrated spacecraft post-landing recovery procedures and the launch vehicle adapter, which will also be used on the uncrewed Exploration Mission-1 (EM-1) in FY 2018, and the crewed Exploration Mission-2 (EM-2) in FY 2021-2022. EFT-1 tested the Orion heat shield at about 85 percent of lunar re-entry velocity, protecting the vehicle from temperatures near 4,000 degrees Fahrenheit. Data collected during the EFT-1 flight test will reduce or eliminate 13 of the top 17 risk drivers for the first crewed flight (EM-2). The flight test also demonstrated nearly 50 percent of the design, development, test, and evaluation (DDT&E) required for EM-2, and included about 50 percent of the software needed for the first crewed mission. Not only did EFT-1 test hardware and software, but it also tested key processes which will be needed for EM-2.

## ORION AND SLS: TRAVELING BEYOND LEO

The dedicated NASA-industry team, working across the nation utilizing all of the NASA Centers and our primary industry partners, Lockheed Martin, Boeing, ATK, and Aerojet-Rocketdyne, is making excellent progress toward developing the next capabilities for human and robotic space exploration missions beyond LEO. The first uncrewed launch of Orion on SLS on EM-1 is slated for FY 2018, and the first crewed launch of Orion and SLS on EM-2 for FY 2021-2022. Both Orion and SLS are being designed to enable multiple missions and destinations rather than being optimized for one particular mission or architecture. Early missions will use cis-lunar space as a proving ground to test out capabilities and operations needed for missions to Mars in a safe and sustainable manner. One early demonstration mission in cis-lunar space will be to rendezvous with and return samples from a near-Earth asteroid. SLS is evolvable to provide progressively greater lift capability, and, with Orion, will enable humankind to successfully navigate the proving ground of deep space, ultimately sending humans to a variety of destinations in the solar system, including Mars.

The Orion spacecraft will be capable of taking humans farther into space than ever before, to multiple destinations as needed, and sustaining them in this challenging environment for longer than ever before. The Orion spacecraft includes both crew and service modules, and a Launch Abort System that will provide for improved crew abort capability. Orion can fly a crew of up to four for 21 days. Orion has a focused and rigorous step-wise test campaign to validate these capabilities in the challenging deep-space environment.

In 2014, in addition to completing the first Orion capsule for flight on EFT-1, the program has begun manufacturing components for EM-1, forging elements for the primary structure, building avionics kits, and procuring parts with a long lead time. In addition, the program continues to work with the European Space Agency as it develops the Orion service module for EM-1.

In FY 2015, the program will focus on preparing for Orion's first exploration mission, EM-1. EM-1 is an uncrewed test flight to lunar orbit, and will be the first pairing of Orion with the Space Launch System. The multi-day flight will provide the program with data, which, combined with data gained from EFT-1, will validate spacecraft design and operations. The Orion program will continue fabricating the crew module primary structure, and start to assemble secondary structures and mechanisms such as propulsion systems, and environmental control and life support. It will also complete its series of parachute tests and begin testing spacecraft avionics. Finally, the program will complete key programmatic reviews, ensuring Orion's readiness to progress to the next phase of the development life-cycle, at which point, the Agency commitment to cost and schedule milestones will be established.

The SLS is a heavy-lift, exploration-class launch vehicle that will transport Orion, as well as cargo and other systems, with a range of lift capabilities from 70 metric tons to LEO, evolving to 105 metric tons and eventually up to 130 metric tons, based on future mission requirements. The evolution of the SLS lift capability fulfills specific, important roles within the exploration architecture, with the 130-metric-ton vehicle enabling future crewed missions to the Mars vicinity.

In 2014, NASA has made significant strides in SLS development. With the conclusion of negotiations for the Stages and Interim Cryogenic Propulsion Stage contracts in June and October 2014, respectively, all major elements for the SLS Block 1 are on contract. For the SLS Block 1 configuration to be flown on EM-1, the program successfully cleared the Agency's Key Decision Point C (KDP-C) milestone in August 2014, marking the transition from program formulation into development. This milestone provides a development cost baseline for the 70-metric ton version of the SLS of $7.021 billion from February 2014 through the first launch and a launch readiness schedule based on an initial SLS flight no later than November 2018. The establishment of this Agency commitment to cost and schedule for SLS at the 70-percent confidence level represents the Agency's formal SLS launch readiness commitment, established after a thorough Agency review process, and addresses Government Accountability Office (GAO) recommendations about matching resources to program requirements. NASA continues to hold the SLS program to an internal launch readiness planning schedule that is earlier than the commitment date. NASA is in the process of updating this internal planning schedule based on a better understanding of work and budget phasing. The appropriations that SLS has received to date have enabled the program to effectively manage risks to the internal planning date.

Both the Core Stage and Booster elements completed their Critical Design Reviews (CDRs) in July and August 2014, respectively, which keeps the program on track for the program-level CDR in 2015. Preparation for next year's CDR has been supported by a wide range of important hardware testing across NASA, including the last buffet wind tunnel test at Langley Research Center, testing of the Core Stage flight computers during the spring, and acoustic model testing of SLS during launch at Marshall Space Flight Center (MSFC) throughout the year.

Production of the first pieces of test and actual EM-1 flight hardware is also underway. The structural rings and engine barrel section for the EM-1 Core Stage are being welded at the Michoud Assembly Facility (MAF), and the first complete verification tank has been completed. NASA is utilizing new, state-of-the-art welding tools to increase the efficiency of Core Stage manufacturing compared to processes used to build Space Shuttle External Tanks. The largest of these tools, the Vertical Assembly Center (VAC) tool, was activated at MAF in Louisiana in July 2014, and the first full-duration test weld on the VAC took place in September 2014. The VAC is the final of six major weld tools at MAF that will produce the SLS Core Stage structure using less than half the labor of Space Shuttle External Tank production. The VAC is scheduled to be validated for flight welds by February 2015 and to be ready to support flight hardware manufacturing. Other major SLS facility work in 2014 includes the new structural test stands at MSFC, which broke ground in August 2014, and the B-2 test stand at the Stennis Space Center (SSC), which, through the use of an innovative Center contracting vehicle, continues to run on schedule and on cost in preparation for the start of Core Stage testing in FY 2017.

Progress continues on the other elements of SLS as well, reflecting both the challenges of advanced spaceflight hardware and the resiliency and innovation of the Government and contractor teams to overcome these challenges. The first SLS flight hardware, the Multipurpose Crew Vehicle Stage Adaptor (MSA), was completed in May 2014 for flight on EFT-1. At SSC, preparations continue to install and test an RS-25 engine in the A-1 test stand, the first time one of these engines (previously known as the Space Shuttle Main Engine) has been tested in over five years. The first hotfire of the RS-25 should take place in early 2015. For the Booster element, a complex chemical and mechanical interaction between the propellant and a new case liner resulted in unexpected voids that posed a testing risk; through an intensive focused effort by the team, new production processes were introduced and a new aft segment, PSA-2, is void-free and ready to support Qualification Motor 1 (QM-1) testing in Spring 2015. SLS remains on track for its program CDR next year.

The Ground Systems Development and Operations (GSDO) team at Kennedy Space Center (KSC) continues to make significant progress on the necessary Exploration Ground Systems (EGS) infrastructure design, development, and refurbishment to support SLS and Orion. KSC also is providing valuable operations expertise to the SLS and Orion teams to address operational issues in the design in order to help reduce eventual production and operations costs. This is a key aspect of assuring long-term sustainability for deep-space human exploration. In 2014, construction of new platforms in the Vehicle Assembly Building at KSC will enable SLS and Orion stacking and preflight processing as planned. Refurbishment and upgrades to a crawler-transporter, which will accommodate up to the 130-metric-ton version of SLS – a vehicle more powerful than the Saturn V – are being performed to support the FY 2018 EM-1 flight of SLS and Orion.

Orion, SLS, and EGS teams are using the latest in systems and manufacturing technology with the intent of developing the safe, affordable, and sustainable systems this country needs to extend human presence to Mars. For example, the Orion team is using time-triggered Ethernet and is taking advantage of the standards for this technology that are used in the automotive industry. Both Orion and SLS are utilizing friction-stir welding (including on large structures, such as the SLS Core Stage), culminating in the most advanced and largest friction-stir weld machine in the world. The EGS team has stripped out the old copper cables from Pad 39B and replaced them with the latest in fiber optics. These are three simple examples of how NASA's Exploration Systems are utilizing and advancing the latest in technology.

In developing the Orion, SLS, and EGS, NASA is seeking to build a sustainable national capability for the long-term human exploration of space. NASA is keeping each element of the program – SLS, ground systems, and Orion – moving at its best possible speed toward the first integrated launch, optimizing each element effort's schedule while being aware of the overall plan. This is best achieved when each element is allowed to progress on its own schedule, rather than being linked too tightly to the others. When tasks related to EM-1 are completed, the workforce can progress to EM-2. NASA is on a solid path toward an integrated mission and making progress in all three programs every day.

The evolving capabilities of these systems will provide the nation with flexibility over the long term to achieve a variety of goals. As we move further into the solar system to establish footholds in a variety of locations, having such flexibility will be important, as future missions can be built on what our astronauts and robotic probes learn in successive expeditions.

## CONCLUSION

NASA's exploration strategy is designed to pioneer multiple destinations in the solar system. Over time, we will move beyond conducting limited-duration forays and begin to lay the groundwork to establish outposts in cis-lunar space. From there, we can expand human presence in the solar system and to the surface of Mars, and utilize *in situ* resources as we extend the reach of humanity. The key to realizing this goal will be to channel all of the factors that have enabled our space achievements to date in a way that will ensure a sustainable foundation on which future generations can continue to build. With the continued support of Congress, this long-term effort will expand the sphere of human life and activity, and draw upon the pioneering spirit and ingenuity, in the face of the seemingly impossible, that have helped make the United States the exceptional nation that it is.

Mr. Chairman, thank you for the opportunity to appear before you today to provide you with a status of our progress on Orion and SLS. NASA has a robust strategy that extends human presence into the solar system in an affordable and sustainable manner, and Orion and SLS are key initial steps in that strategy. I would be happy to respond to any questions you or the other Members of the Subcommittee may have.

# INDEX

## D



## E



## F



## G



## H



## I



## J



## L